A Simple Guide

to

Verizon iPhone

by
Mary Lett

LUMINIS BOOKS
Published by Luminis Books
1950 East Greyhound Pass, #18, PMB 280,
Carmel, Indiana, 46033, U.S.A.
Copyright © Luminis Books, 2011

PUBLISHER'S NOTICE

Cover art direction and design by Luminis Books.

ISBN-10: 1-935462-45-8
ISBN-13: 978-1-935462-45-3
Printed in the United States of America

10 9 8 7 6 5 4 3 2 1

Simple Guides

give you

Just the Facts

Get up to speed on Verizon iPhone—fast!

Simple Guides: get you started quickly.

No extra clutter, no extra reading.

Learn how to set up your Verizon iPhone, **customize your screens and ringtones,** take pix and videos, **and download all the coolest apps.**

Learn the nuances of **using Verizon iPhone** to maximize your **business and personal time. Start using voice mail, GPS, and text** set up your iPhone the way you want, and **start having fun!**

Acknowledgments

As with any book, there are always those who help you get through the process and deserve your gratitude. This book is no different. First and foremost is my son John Lett, who cheered me on. My good friends Debbi Mitchell, Penny Wright, and Andrea Taylor for always listening. My good friend Debbie Abshier for signing me on to this project. And last but not least Chris Katsaropoulos and Kelly Henthorne, editors extraordinaire, who not only edited this book, but showed me the ropes. A great big thanks to one and all, as I would never have made it without you!

Table of Contents

Chapter 1

Things to Do Prior to Activating Your New Verizon iPhone

Congratulations! You have purchased the new Verizon iPhone 4, one of the most anticipated products in some time. What are the key things that you need to do prior to activating and properly setting up your new iPhone?

The biggest issue with acquiring a new phone is transferring important information from your old phone to the new one. In this chapter, we cover the best ways to preserve your current contact information, music, videos, and other important data that you want to save.

Additionally, we give those of you who already have an iPhone some information on how to change to a Verizon iPhone.

Four Key Steps to Take Prior to Activating Your iPhone

Here are the four key steps that you should take prior to turning on your new phone:

➢ Archiving your current voice mail
➢ Downloading Apple iTunes
➢ Preparing to transfer your contacts

> ➤ Transferring your pictures, videos, music, and documents from your current phone

Archiving Your Current Voice Mail

After your new iPhone is activated, your current voice mail will be deleted unless you subscribe to Visual Voice Mail, which is a service that Verizon Wireless provides for $2.99 per month. Contact Verizon Wireless to add this service to your contract. If you have Visual Voice Mail, complete the following steps to archive your current voice mail:

1. Click **Message** under the Visual Voice Mail program.
2. Click **Options.**
3. Click **Save a copy/Archive.**
4. Click **Save to SD Card or External Memory.**

If you do not subscribe to Visual Voice Mail, you will need to write down any voice mail messages that you want to keep.

Downloading iTunes

One of the most important steps in setting up your new iPhone is going to the Apple iTunes website and downloading the Apple iTunes application. You cannot turn on your new iPhone until you have this program. You use this program to download and manage your data and applications between your iPhone and your computer. Without it, you cannot download apps (you learn how to get and use apps in Chapter 4), music, movies, or podcasts or have the ability to ping. iTunes will become the lifeblood of your new iPhone, enabling you to do all of the great things that you have been looking forward to doing.

To download the Apple iTunes application, go to www.apple.com/itunes/download/. At this site, complete the following steps:

1. Click the **download** icon to begin downloading.

Figure 1-1: Apple iTunes Download Page

2. Click **Yes** to accept the terms of Apple's licensing agreement in order to continue with the download.
3. Click **Yes** to save the file to your computer.
4. The download itself takes about 5 minutes, depending on the speed of your Internet connection.

After downloading and installing iTunes, you need to set up an Apple Store account:

1. At www.apple.com/itunes, click the Store menu on the top left side of the iTunes window and then click **Create Account.**

Figure 1-2: Apple iTunes Create Account Menu

2. Click **Continue** at the bottom right side of the screen.
3. Fill in the form with your e-mail address, password, password question, and birth date.
4. Click **Continue** to go to the next screen.
5. Fill in the next form with your credit card information and mailing address.

6. Click **Continue** to go to the next screen. You will get a message telling you that you will receive an e-mail to complete the registration process.
7. When you receive the e-mail, click on **Verify Now** to complete the registration process.

Dear Your Name,

You've entered <u><your</u> email address> as the contact email address for your Apple ID. To complete the process, we just need to verify that this email address belongs to you. Simply click the link below and sign in using your Apple ID and password.

<u>Verify Now ></u>

Wondering why you got this email?
It's sent when someone adds or changes a contact email address for an Apple ID account. If you didn't do this, don't worry. Your email address cannot be used as a contact address for an Apple ID without your verification.

For more information, see our <u>frequently asked questions</u>.

Thanks,
Apple Customer Support

Figure 1-3: Apple Verify E-Mail Address

8. Next, fill in your login name (e-mail address) and password.

9.   Your registration is now complete. You can go to the iTunes store and purchase apps (see Chapter 4) or music.

After setting up your Apple Store account, you may also want to go to the iTunes page and review the short tutorials on how to use the various iTunes features. Each of these tutorials lasts one to three minutes and provides a great way to get up to speed quickly with the various aspects of iTunes.

Transferring Your Contacts

The next important step in preparing to use your new iPhone is transferring your existing contact data. Taking care to transfer your data properly will help you avoid spending a lot of extra time re-entering contacts later.

Verizon offers a program called Backup Assistant to help you back up and transfer your existing contacts. You can download this program to your current phone in order to back up the contacts you already have. Then you can install Backup Assistant on your new iPhone to transfer your contacts to the new phone. If you are currently using a Windows phone, you can download the Windows Device Backup Program to transfer your contacts to Outlook.

To download the Backup Assistant or Windows Device Backup Program, go to http://www.verizonwireless.com and log into your personal account.

1.   Click **My Services** and then click **Backup Assistant**.
2.   Click the **Backup Assistant** and then the **Download** link. After you have downloaded either of these programs, your phone can be synced.

3. Click on the **Sync** link on the Backup Assistant, and your phone will copy your contacts to your computer.

After the phone is synced, your data is ready to transfer to your new iPhone.

Transferring Pictures, Videos, Music, and Documents

Beyond contacts and voice mail, you probably have photos, videos, music, and documents that you want to move to your new iPhone.

To transfer these items from your current phone to your new iPhone, do the following :

1. Log into your Verizon wireless account.
2. Click **Media** and then click **V Cast Media Manager**.

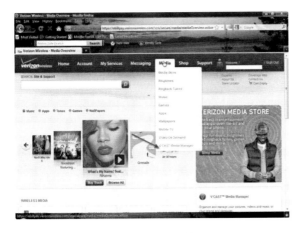

Figure 1-4: Verizon Wireless Media Menu

You can download a free copy of this program, which enables you to transfer these items to your computer. You can also subscribe to a service starting at $2.99 per month to have 25GB of online storage capacity. The advantage of online storage is that it frees storage space on your phone. If you choose this option, part of the download will be an app for both your phone and your computer to gain quick access to your online storage files. Click on the download button underneath the type of storage that you prefer, and the program will be downloaded to your computer to manage the various forms of media.

Transferring Your Current iPhone to Verizon

Okay, first the bad news: It is currently not possible to transfer your current iPhone to a Verizon iPhone account. However, Verizon does offer a trade-in program in which they will give you a predetermined credit for your current device. This credit then can be used to purchase a Verizon iPhone.

Go to http://www.verizonwireless.com and click on the **Trade-In** program link. When you are at the Trade In page, click **Appraise Your Device.** Choose your manufacturer and model and answer the three yes/no questions. Next click **Continue** on the right side of this page.

Figure 1-5: Verizon Wireless Appraise Your Device Page

Verizon will give you the value for your current phone, which can be used toward the purchase of the Verizon iPhone.

Following these steps to properly set up your new iPhone and to transfer your current data will save you a lot of time when you start using your new iPhone. After you have set up the new iPhone and ensured that the data from your current phone is secure, you are ready to turn on your new iPhone and start having fun with it!

Just the Facts About Things to Do Prior to Activating Your Phone

➢ You must write down your current voice mail messages or subscribe to Visual Voice Mail for $2.99 per month to save your current messages.

9

- ➤ After you activate your iPhone, you will no longer be able to access your old voice mailbox.
- ➤ Downloading Apple iTunes and setting up an iTunes account is a must.
- ➤ Without an iTunes account, you cannot activate your iPhone.
- ➤ Your iTunes account will allow you to download and manage apps, music, movies, TV shows, podcasts, audiobooks, and Ping.
- ➤ Verizon Backup Assistant or Windows Device Backup Program needs to be downloaded from Verizon Wireless to back up your contacts to transfer them onto your iPhone.
- ➤ After you download Backup Assistant or Device Backup, you must sync your old phone to save your contacts to your computer.
- ➤ You can upload your synced contacts to your new iPhone.
- ➤ Pictures, videos, music, and documents can be saved by downloading the free V Cast Media Manager or by subscribing to the V Cast Media Manager online storage service.
- ➤ If you already have a non-Verizon iPhone, you cannot change your service to Verizon Wireless. You will need to buy a Verizon iPhone.
- ➤ Verizon Wireless has a phone trade-in program for people who have other cell phone carriers, but who would like to change to a Verizon iPhone.

Chapter 2

Setting Up Your New iPhone and Using Basic Functions

After you have completed all of the pre-activation steps covered in Chapter 1, it is time to turn on your new iPhone and set it up for use.

Initial Set Up of Your New iPhone

Your new iPhone comes with the following items:

- ➢ Set of earphones
- ➢ USB cable
- ➢ USB power adapter

The button to turn your iPhone on and off is located on the top of the phone on the opposite end from your USB Connection (see Figure 2-1). To turn your iPhone on, press the on/off button on the top of your iPhone. When the screen opens, slide your finger across the arrow to unlock. Your home page appears after you unlock your iPhone.

To turn your iPhone off, press the on/off button on the top of the iPhone (see Figure 2-1). A red arrow shows up on your home screen, prompting you to slide your finger across the screen to turn your iPhone off. Press **Cancel** at the bottom of the screen if you change your mind and do not want to turn your phone off.

Figure 2-1: Your New Verizon iPhone

Activating Your iPhone

Your iPhone comes with a partial charge. To activate your new iPhone, you must complete the following steps:

1. Plug the USB cable into your phone.
2. Plug the USB cable into your computer.
3. The iPhones' Activation software automatically loads on to your computer. Your iTunes program automatically opens to the Verizon Terms and Conditions page.

Figure 2-2: Verizon Terms and Conditions Page

4. Click **I Accept** (see Figure 2-2) after you have read the Verizon Terms and Conditions page. A message appears on your iPhone telling you that your iPhone is activating. This takes a couple of minutes.

5. Click **Continue** when you are prompted to register your iPhone.

Figure 2-3: Verizon Let's Get Started Page

6. Click **I Have Read and Agree to the iPhone Software License Agreement** .
7. After you have read the agreement, click **Continue.** You will again see the Terms and Conditions page.

Figure 2-4: iPhone Software License Agreement Page

14

Your activation and registration is now complete. (If you go to www.Verizonwireless.com and log into your account, your iPhone will show as your current device. This lets you know that your iPhone is now registered with Verizon Wireless.)

Locking Your iPhone

After one minute of being idle, your iPhone automatically locks, and the screen goes dark. Later in this chapter, I discuss settings, and you learn how to change this setting if you would like your phone to stay open longer.

Why is locking your phone so important? That's simple: When locked, your phone can never call someone by mistake when its being jostled in your pocket, purse, or backpack.

To unlock your iPhone, press the **Home** button on the front of your phone (see Figure 2-5) or the on/off button (see Figure 2-1) to turn on the screen. In either case, just above the Home button is a prompt that goes along the bottom of the screen that tells you to Slide to Unlock.

Figure 2-5: iPhone Home Button

Run your finger from the left side of the screen to the right side of the screen to unlock your iPhone. After the iPhone is unlocked, your Home screen once again appears.

Using the Multi-Touch Display

All of the screens and functions of your new iPhone are manipulated by using your fingers. No stylus is attached to the Verizon iPhone, and even if you tried to use a stylus, it would not work on the iPhone. Fingernails do not work either. There are four ways to use your fingers to work the iPhone:

➢ Tap
➢ Flick
➢ Pinch and spread
➢ Drag

Tapping

The most frequently used type of finger touch is a **tap**. You use taps to open apps (more information on apps in Chapter 4). You also can use tapping to choose songs, photos, movies, or podcasts.

You can use taps to access the various keyboards to send e-mails, texts, or phone calls. As the keyboards take a little getting used to, you might want to practice writing notes using the Notes icon, which comes on your new iPhone. Tapping is just like how it sounds: You tap the tip of your finger over a key on the keyboard or keypad or over an icon. Tapping an icon once opens it for you. Tapping an alpha or numeric key types the character.

Double-Tapping

Double-tapping (tapping twice in rapid succession) is used to zoom in or out of Web pages, maps, and e-mails.

While using the tap method to type, if you type an incorrect key, touch your finger to the screen, hold it there for a few seconds, and a little magnifying glass appears. The magnifying glass turns into a message such as cut, copy, or select to help you do some editing to your message.

Figure 2-6: iPhone Magnifying Glass

Flicking

Flicking is used to scroll through your list of contacts, text messages, song lists, or any other list. To flick, put your finger down and roll the list up, down, or sideways. Flick your finger, and the list scrolls. This way you can quickly see what is in any list.

You can practice flicking on your home page. The initial setup of your iPhone has information on two pages, which roll from side to side. You can flick to move back and forth between the two pages. This motion is also used to unlock your iPhone or to turn it off.

Pinching and Spreading

You use pinching and spreading to expand your image. For example, to access the map program that comes with your new phone tap on the map icon. The map opens to a world map. Place two fingers in the middle of your screen. While continuing to keep your finger on the screen, spread your fingers toward the opposite corners of the screen. The map shows more detail. If you move your fingers off of the screen and repeat the preceding movement, you begin to see more and more detail as you continue to spread your fingers. You eventually wind up with the satellite view of a street. Conversely, to go back to the original view of the map, place your fingers in the corners of the screen and pinch them together in the middle. The map has less and less detail the more times you pinch your fingers together.

Dragging

The last type of finger movement that you need to use is dragging. Gently place your finger on the screen over the icon that you want to move. Do not tap the screen, or the icon opens. Hold your finger over the icon for a few seconds. Note that the icons start to wiggle. You can drag the icon that you want to move with your finger to wherever you want to move it. After the icon is moved, the icons on your Start menu continue to wiggle for just a bit. If this bothers you, you can tap the Home button, and the icons once again lock into place.

Charging Your Phone

Use the USB power adaptor to charge your phone. The USB power adaptor can be used as a USB plug attached to your computer, or it can be plugged into the power adaptor and into a regular electrical outlet. When you tap on the Home button, a large battery is pictured, which shows you when your battery is charged to 100%. You can still use your iPhone while it is charging. The charge lasts approximately seven hours depending on how you use your phone. Power can be saved by turning off the WIFI if you are not using it. To turn off the WIFI

1. Click **Settings**.
2. Click **WIFI**.
3. Click **Off**.

You can also shut down Bluetooth if you are not using a Bluetooth headphone. To do this

1. Click **Settings**.
2. Click **General**.
3. Click **Bluetooth**.
4. Click **Off**.

Another feature that will eat up your charge is the GPS.

If you do not need or want your phone to track your location, turn it off. To turn it off, perform the following steps:

1. Click **Settings**.
2. Click **General**.
3. Click **Location Services**.
4. Click **Off**.

A charge percentage is available on the monitor that you can turn on using the following steps:

1. Click **Settings**.

2. Click **General**.
3. Click **Usage**.
4. Click **Battery Percentage**.

After this feature is turned on, a percentage shows just to the left of your battery gauge on the top right side of the screen.

Figure 2-7: Battery Charge Percentage

Screen Function Basics

Changing the look of your new iPhone is just a tap away. In seconds you can change your sound settings, screen brightness, your wallpaper, general settings, and about information.

Figure 2-8: Setting App

To make any of these changes, tap on the Settings app. The following sections take you through the screen function basics:

Sounds

Use the following steps to change sound settings:

1. Tap **Sounds**.
2. Tap the **On** button if you want to turn vibrate off.
3. Tap the volume line if you want to increase the volume of your phone.
4. Tap **Ringtone** to change the ringtone of your phone. You can choose from several tones. If you tap on the various ringtones, you get a sample of what that tone sounds like.

5. **Ta**p **Text Tone** to change the tone for receiving and sending texts. Again, you can choose from sever**al options**. **Simply tap** on the name of the tone to get a sample of the sound.

Several other sound choices are all yes/no questions based on your individual preference. An example of these is **New Voicemail**. The question is, do you want a tone when you receive a new Voicemail, Yes or No? The default is yes. To change the tone, tap on the **Yes**, and it will become No. Review the other items in this category to make sure they are set up the way you want them to be.

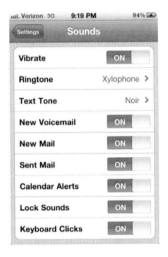

Figure 2-9: Sound Settings

Screen Brightness

When brightness is on, it adjusts your screen for the various lighting around you. It should darken outside and lighten when you are inside. By moving the lever to the left or right, you increase or decrease the brightness of your screen. This setting uses battery power. The brighter the screen, the more battery power you use. I suggest that you leave the auto brightness on so that the phone adjusts the brightness to your lighting conditions. This is the best choice for your battery power conservation.

Figure 2-10: Screen Brightness Bar

Wallpaper

Your phone comes with a default gray background. To change this to something different, perform the following steps:

1. Tap the **Settings** app.
2. Tap **Wallpaper**.
3. Your current screen is pictured; tap on that picture.
4. Choose between Wallpaper (preset pictures) or Camera Roll (your pictures).

5. Tap any of the images that interest you. A preview screen comes up.

Figure 2-11: Wallpaper Preview Screen

6. Tap **Cancel** to go back to your other choices or **Set** to choose this option.
7. If you choose Set, the next step is to choose whether you want this option for your Lock screen (initial screen when you press the Home button), Home screen, or both.
8. After you set your options, you are taken back to the Wallpaper or Camera Roll screen.
9. Press the Home button to go back to your Home page.

General

Under General Settings are a variety of settings that are preset to optimize your ease of using your phone. Following is a general overview of the various General Settings. You can choose whether you want to go in and make any changes to your options.

1. **About** — About gives you general information about your phone. It tells you the following things:
    - ➢ Name of your network
    - ➢ How many songs, videos, photos, and apps are stored on your phone
    - ➢ Storage capability and availability
    - ➢ Software version being used
    - ➢ Wireless carrier
    - ➢ Phone model, serial number, and WI-FI and Bluetooth addresses
    - ➢ Your smart card identity number, MEID (Mobile Equipment Identity)
    - ➢ Modern firmware or the version of the cellular transmitter
    - ➢ Legal and Regulatory, all of the legal and regulatory information regarding your iPhone

2. **Usage** -- Usage gives you more information about your phone. These items are
    - ➢ Battery percentage (see "Charging Your Phone")
    - ➢ Time since you last charged your phone
    - ➢ Length of your calls
    - ➢ Cellular Network Data, the amount of data you have sent and received
    - ➢ Tether Data, how much data has been sent or received from a device that is tethered to your phone such as a personal computer
    - ➢ Reset Statistics option, which can be tapped to erase all of the preceding statistics

Making a Phone Call

Now that your phone is all set up and ready to go, it is time to start using it to make phone calls. Tap the **Phone** icon on the Home screen.

Figure 2-12: The Phone Screen

Along the bottom of the screen you see the following options:

➢ Favorites
➢ Recents
➢ Contacts
➢ Keypad
➢ Voicemail

Keypad

The top of the screen is the keypad, which disappears when you choose any of the options other than Keypad.

By tapping on Keypad, the keypad will return.

To make a basic phone call when you know the telephone number, use the Keypad to type the number, then tap **Call,** and the phone dials. If you type an incorrect number, tap on the **X** and the number is erased.

Favorites, Recents, and Contacts all allow you to access numbers that are stored on your phone in various forms.

Next we explore making calls using these options.

Contacts

Contacts are your personal phone book. Contacts can be added through your computer (this is covered in Chapter 5, "Syncing Your Phone") or directly through your iPhone.

On your iPhone, you can add Contacts in one of two ways: either through the Contacts App, which can be found on your Home page, or through the Task menu under the Phone icon. Either choice allows you to add Contacts and use them in the same way. To add Contacts through your iPhone and then use them to make a phone call, use the following steps:

1. Tap **Contacts** (either the app on the Home page or the menu choice under the Phone icon).
2. Tap on the **+** to the right of All Contacts at the top of the Contacts page.

3. The Contacts page allows you to add the person's (or business') name, phone number, e-mail address, mailing address, and/or some miscellaneous fields.

    ➤ Tap on the field (name, phone number, etc.) that you want to change and begin typing.

    ➤ Any field in blue can be tapped to either change the label or add information under that category. (That is, Phone defaults to Mobile. Tap Mobile, and an entire menu of options opens.). You can type information in any of these additional fields, and they will be placed on the person's Contact page. You can also use this to swap fields. Whatever fields you complete will either replace the original field or add fields.

    ➤ You can add a photo from your photo album or by taking a photo of the person by tapping the Add photo box.

    ➤ Ringtones can be added to distinguish one contact from another or to highlight specific contacts.

    ➤ After all of the information that you want to include is completed, tap Done, and the contact is saved.

4. Now that your contacts are added, you can tap **Contacts** to see an alphabetical list of your contacts. You can flick to search through your list to find a particular contact. When you have found your contact, tap on the name. That person's Contact page opens up.

5. Tap the number you want to call, and the phone begins dialing.

6. While you are talking, your screen goes black.

7. When your call is complete, a screen prompts you to end the call.

8. Tap the **End call** prompt, and the call is ended.

Favorites

Quick options and often used numbers are stored in Favorites. This is your speed dialing option. To set up a contact as a favorite, tap **Contact**. When the Contacts page is open, tap **Add to Favorites**. You are prompted to add the person as a voice or face-to-face call. After the option is tapped, the contact automatically is added to your Favorites page.

To remove someone from your Favorites page, complete the following steps:

1. Tap **Edit** located to the left of Favorites at the top of the page.
2. Tap the **red circle** to remove someone from your Favorites list.
3. When you are finished removing names, tap **Done,** and your changes are saved.

After your Favorites are added, tap on the person's name, and your phone dials automatically. When the call is completed, you are prompted to end the call. Tap on this prompt, and your call is ended.

Recents

Recents is a list of your most recent calls. At the top of the Recents page, you have the option of looking at all of your calls or just the missed calls. You also have the option to clear all of your recent calls.

To make a call using Recents, tap the person's name or the phone number, and your phone begins to dial.

Voicemail

The first time you tap on the Voicemail menu, you need to set up your voicemail greeting:

1. Next to Voicemail at the top of the page is Greeting. Tap **Greeting.** Your choices are Default or Custom. Default is the canned message. Custom enables you to record your own greeting.

Figure 2-13: Voicemail Greeting Screen

2. Tap **Record** to start recording your greeting.
3. When you finish recording, tap **Save** to store your message.

A red circle with a number indicates how many missed calls or voicemails you have. To get your voicemail or missed calls, do the following:

1. Tap the **Phone** app. A red circle with a number is shown on the Recents and/or the Voicemail menus.
2. To see who has called or left a message, tap the menu item; missed calls appear in red on the Recents menu, and new Voicemail messages have a blue dot on the left side.
3. To return the missed call, tap the number or name, and the phone dials the number.
4. To listen to the voicemail, tap the missed message, and it begins to play.
5. After you listen to the voicemail, you can either do nothing, tap **Call Back** to return the call, or **Delete** to erase the message. If you delete the message, a Delete Message line is added to your voicemail list.
6. Tap **Delete Message** and then **Clear Messages,** and all deleted messages are removed from your phone.

Voice Dialing a Phone Call

Your phone can also be dialed with voice signals. To do this, press down on the Home button. After a few seconds, the Voice Control screen appears. Simply state the number that you want to call or the person's name, and the phone begins to dial. Continue the call as explained previously. If you are calling a person on your contact list and they have more than one number listed, a recorded voice asks you which number you would like to use.

Receiving a Phone Call

When your phone rings, a screen appears that shows the contact or number that is calling. The screen also shows two choices, Answer or Ignore. You can tap **Answer** to take the phone call;

Ignore to not be bothered; or nothing, which is the same as Ignore. If you answer the phone, the screen goes black until the call is complete. When the call is complete, another screen appears with a prompt to end the call. Tap on this prompt, and the call ends.

Figure 2-14: iPhone Call Screen

While you are on a call, a Call screen appears. You are given the following seven options with this screen.

Mute

Tap **Mute** to block the other person from hearing what you are saying or to block background noise.

Keypad

Keypad is used to call additional numbers or respond to an automated menu system.

Speaker

Speaker activates the speaker phone so that you can talk hands free.

Add Call

Add Call has many options. You can put one call on hold and answer the second call. When you receive a call, a screen comes up giving the option to Ignore, Hold Call & Answer, or End Call and Answer. Tap the option you prefer when dealing with the incoming call.

Use Add Call to set up a conference call. You can have up to five people on a conference call. To set up a conference call, take the following steps:

1. Call the first person.
2. After the person answers, put her on hold.
3. Tap **Add Call** and dial the second person.
4. When he answers, tap **Merge** and the second person joins the call.
5. To add others to the call, repeat steps 2–4.
6. To drop a caller from the conference, tap **Conference** and then tap the red circle with the phone in it that appears next to the call. Tapping **End Call** removes the caller from the conference.
7. To speak privately to someone in the conference, tap **Conference**, then tap Private next to the caller with whom you want to talk. When you are finished with the side conversation, tap **Merge Call** and everyone is back in the conference call.

8. You can add a new incoming caller into the conference by tapping **Hold Call + Answer** followed by **Merge Calls**.

Hold/Face Time

Tap **Hold** to pause a call for a few minutes.

Face Time calling is a live Wi-Fi feed of the person to whom you are talking. In order to have one of these calls, both persons must have an iPhone 4 and a Wi-Fi connection. To access Face Time, tap **Face Time**. The other person receives an Accept or Decline screen. If they accept your call, you begin to see them on your screen.

Depending on how much background you want the other person to see, you can choose either the front or main camera to do the videocasting. For a narrow view use the front camera. For a wider view, use the main camera. To switch the camera being used, tap the camera on the right side of the tool bar.

If you never want to use Face Time, tap **Settings**, tap **Phone**, and tap the **On/Off** option next to Face Time to turn it off.

While on a Face Time call, you can access other information on your phone by tapping your **Home** button. You will not be able to see the other person while you are getting whatever information you want. To continue seeing the other person, tap the green bar at the top of the screen.

Contacts

Contacts enables you to look up an additional contact while you are on the phone with someone.

End Call

The final option that appears, End Call, was discussed earlier in this chapter.

Using a Bluetooth Device

To set up your Bluetooth, tap **Settings**, **General**, **Bluetooth**. A Bluetooth symbol is now placed on the status bar at the top of your phone.

Figure 2-15: iPhone Status Bar with Bluetooth Symbol

If the symbol is blue or white, your phone is communicating with your Bluetooth device. If it is gray, Bluetooth is turned on, but it is not communicating with your Bluetooth devices. To pair your device, see the instructions that came with your device to pair it.

To remove a Bluetooth device from your phone, tap on the arrow to the right of the device name. If you want to remove it temporarily, tap **Connected** next to the device name.

You should see Not Connected. To restart the connection, tap **Not Connected**.

Using the Internet

To access the Internet, tap the Safari icon in the Task menu. You then can choose which program you want for your Home page. To choose a page, search for the page in the search window and then tap the arrow coming out of the box above the Home button. Your page is saved as your Home page when you open Safari.

Figure 2-16: iPhone Safari Screen

The tool bar at the bottom of the screen allows you to do the following things:

➢ Scroll through recently opened web pages using the arrows.
➢ Add bookmarks, choose your Home page, and open a mail link by using the arrow coming out of the box.
➢ See a list of your bookmarked pages by using the open book icon.
➢ Review the number of pages that you have surfed and what they are by using the page counter.

Web pages on the iPhone work exactly the same way that they do on your home computer. Since on your phone the web sites are very small, you can tap on the screen to zoom in on a certain area of the site. You can also turn your phone sideways to widen the view, thereby making the site bigger. In this view, you have to scroll down the page to see the entire page. You can still tap on the screen to zoom in on a particular area of the web site.

Receiving and Sending E-Mail

When you activated your iPhone, your e-mail account information was automatically loaded into your phone from your computer. To check what was loaded onto your phone, tap **Settings**, **Mail**, **Contacts**, **Calendars**, and then your e-mail server. You should see all of your account information.

Receiving E-mail

To receive e-mail, tap the **Mail** icon on the Start menu. Any new e-mail loads after the icon opens. You can flick the e-mail list up and down to see what e-mail you have received. To read an e-mail, tap on the e-mail, and it opens up. After you have read the e-mail, you have the following options:

➢ File the message in another folder by tapping the file folder in the tool bar. When you tap the file folder in the tool bar, a list of your folders appears (more about folders in Chapter 3). Tap on the folder where you want the message to be filed. The file moves to that folder.

➢ Tap the trash can to delete the message. It is moved to the Trash mailbox.

➢ The arrow to the right of the Trash can is the Reply/Forward button. Tap this button to reply or forward the message.

➢ To start a new message, tap the New Message box in the right corner of the tool bar. A New Message screen appears including a keyboard. The plus sign on the To: line is used to access your contacts. Tap the plus sign, and your contact list appears. Tap the contact you want to e-mail. When your message is ready to go, tap Send, and your message is sent. If you are unable to send a message due to an error with your e-mail account, contact Apple Cares within the first 90 days of owning your phone. You can contact Apple Cares at 1-800-694-7466 or online at www.apple.com/contact.

E-mails can be threaded, or grouped, by a particular category. To set up a thread:

1. Tap **Settings**.
2. Tap **Mail**, **Contact**, **Calendars.**
3. Flick the list to find **Organize By Thread**.
4. Make sure the On/Off button is in the On position. If it is on, all like messages should be grouped together.

A search inbox window is at the top of your e-mail list. To quickly search through the e-mail, type a search word in this window, and all relevant e-mails will appear.

Other E-Mail Options

Other items of note for using e-mail are as follows:

➢ To open an attachment in an e-mail, tap the attachment, and it should open.
➢ To see all recipients of an e-mail, tap Details to the right of the senders' name.
➢ To add an e-mail sender to your contact list, tap the person's name or e-mail address and then tap either Create New Contact or Add to Existing Contact.
➢ Messages can be marked unread by tapping Mark as unread, which can be found near the top of the e-mail in blue with a blue dot.
➢ To expand or reduce the size of your e-mail, use pinch and spread.
➢ If you tap a link in the e-mail, Safari will open and display the website.
➢ To save an outgoing e-mail to drafts, write your message and tap Cancel in the upper left corner. Options appear. Tap the Save Draft option. The message is saved in a Drafts mailbox, which is located on the All Mailboxes page.

More information about e-mails is provided in the next chapter.

Your phone is ready to use, and you have learned about the basics of using it. In the following four chapters, you learn more on how to use the various functions and apps, how to sync your phone, and finally some troubleshooting tips.

Just the Facts About Set Up and Basic Functions

➢ Your phone comes with a set of earphones, USB cable, and a USB power adapter.

➢ To activate your phone, connect the USB cable to your phone and then to your computer. The software automatically loads and activates your phone after you agree to Verizon Wireless' terms and conditions.

➢ Your phone automatically locks after one minute unless you go into Settings and extend the time lapse.

➢ You can move around the Multi Touch display of your new iPhone in four ways: tapping, flicking, pinch and spread, and drag.

➢ To charge your phone, connect the USB cord to the USB power adaptor. Connect the USB cord to your phone and then to an electrical outlet. It takes about an hour to charge.

➢ The charge lasts around seven hours depending on your usage.

➢ You can open the Battery Percentage meter under the Settings app to monitor your battery usage.

➢ Your new phone can be customized by changing the Sounds, Screen Brightness, and Wallpaper.

➢ To make a phone call, tap on the phone icon and choose someone from your contacts, someone you recently talked to, or one of your favorite contacts. You also can type the number using the keypad.. Touching the number in your contacts will automatically start your phone call. If you use the keypad, you need to tap Call to start your call.

➢ To add a contact, tap Contacts, tap the + in the upper right corner of your screen, and fill in the information that you want to about the individual or business. When you are finished, tap Done, and your contact is saved.

➤ To add a contact from a recent call, tap the arrow to the right of the phone number. You are taken to the Contacts page. Tap Create New Contact. Complete the information that you want to keep. Tap Done, and your contact is saved.

➤ You can voice dial someone by pressing on the Home button for a few seconds. The Voice Control screen appears, and you can say whom you would like to call.

➤ To receive a call, tap Answer or Ignore on the screen that pops up when the phone rings.

➤ To end a call, tap End Call, which appears at the end of your call.

➤ You can set up a conference call with up to five people.

➤ You can use Face Time to see the person with whom you are talking.

➤ To set up a Bluetooth device, tap Settings, General, Bluetooth. Next pair your Bluetooth device to your iPhone.

➤ Use Safari to access the Internet.

➤ E-mail can be downloaded by tapping on the Mail icon.

➤ You can reply to a received e-mail or start a new e-mail by tapping the New Message box in the right corner of the Mail toolbar.

Chapter 3

Syncing Your New iPhone

After you have backed up your old phone and learned the basics of your new phone, it is time to start restoring your contacts, appointments, and photos. You also might want to load music, movies, podcasts, or other photos from your computer. This chapter explores just how to transfer all of your data.

The Importance of Syncing Your New iPhone

The main reason to back up or sync your new iPhone is to protect your data in case your phone is broken, stolen, or lost. Imagine losing your new phone after loading a thousand songs. If you have your phone synced, your songs are all saved on your computer; replacing them would take only a few easy steps and 10 to 15 minutes. If you do not sync your phone, it could take you days to replace that many songs.

Just the Facts About Syncing Your iPhone

Syncing your iPhone is very similar to the process used to sync an iPod. To sync your iPhone, follow these steps:

1. Connect your iPhone to your computer with the USB cable that came with your iPhone. When you connect your iPhone to your computer, iTunes should automatically load. If it does not, make sure that you are connected directly to your

computer and not through a USB hub. If iTunes still doesn't launch, try launching it manually by clicking the iTunes icon on your computer. If you have pictures on your phone, the first thing that will launch in iTunes is a window asking whether you want to download your photos.

Figure 3-1: iTunes Photo Download Window

2. Select your phone in the iTunes list.
3. Name your phone.
4. You will be prompted to click the items that you want iTunes to automatically sync with your iPhone. Your options include: your contacts, calendars, bookmarks, notes, e-mail, and apps You can click all or selected items.
5. Check **Automatically Sync Applications** if you want iTunes to do this.

NOTE: If you prefer to sync manually, leave the boxes in step 4 and 5 unchecked.

6. To have iTunes launch automatically and sync your phone whenever you connect it to your computer, check the **Open iTunes When This iPhone is Connected** check box in the Options area.

Figure 3-2: iTunes Options Screen

You will not be able to sync your iPhone automatically if the Prevent iPods, iPhones, and iPads from Syncing Automatically option in the Devices window is enabled.

To clear this option so that you can sync automatically, do the following:

1. Click **Edit**, **Preferences**, **Devices**.
2. Click **Prevent iPods, iPhones, and iPads from Syncing Automatically** to remove the check. When the box is cleared, you can go back and click **Sync**; you're ready to go.

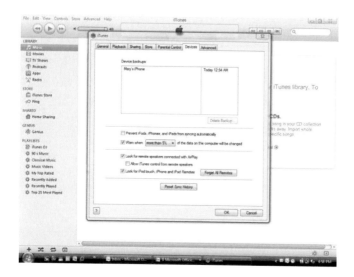

Figure 3-3: iTunes Preference—Device Window

To sync manually, do the following:

1. Uncheck **Open iTunes** when this iPhone is connected.
2. Next click **Sync** to sync manually. This option can be found on iTunes' Summary window (see Figure 3-2).
3. Check **Sync Only Checked Songs and Videos** if you want to sync only items that you have check marks to the left of their names in your iTunes library (see Figure 3-2). We will discuss this process more in Chapter 4 under using your iPod.
4. Click **Convert Higher Bit Rate Songs to 128 kbps AAC** if you want iTunes to automatically create smaller audio files so that you can fit more music on your iPhone (see Figure 3-2).

5. Click **Manually Manage Music and Videos** if you want to turn off automatic syncing of your music and videos. (see Figure 3-2).
6. Click **Encrypt iPhone Backup** if you want to password protect your iPhone backups (see Figure 3-2).

If you've change any sync settings since the last time you synched your phone, instead of saying, "Sync", the button will instead read "Apply."

A Sync in Progress screen appears while your phone is syncing. When the sync is complete, iTunes displays a message stating that the sync is complete and that it is okay to disconnect your iPhone. If you disconnect your iPhone prior to seeing this message, your syncing can fail. Do not disconnect your iPhone from your computer until you see that it is okay to do so.

To cancel a sync, drag the slider on the iPhone (it reads Slide to Cancel) during the sync. If you get a phone call while you are syncing your phone, the sync is cancelled automatically so that you can disconnect your iPhone from your computer and take the call. When finished with your call, reconnect your phone to your computer, and the sync continues.

Manually Syncing Your iPhone

To manually sync your iPhone, you need to tell iTunes how you would like to perform your sync. To do this, go to the Info page, which you can find on the sync window next to the Summary page.

Figure 3-4: iTunes Sync Info Page

The Info page has six different sections:

➢ Mobile Me
➢ Sync Contacts With
➢ Sync Calendars With
➢ Sync Mail Accounts From
➢ Other
➢ Advanced

Mobile Me is discussed in depth in Chapter 4.

Just the Facts About Syncing Contacts

➢ Sync Contacts With determines how you will sync your address book contacts.
➢ To what program do you want to SYNC Contact With? If Outlook is loaded on your computer, it displays as the

48

default program. You can change to a different program by clicking the up or down arrows next to Outlook in iTunes under the info tab. iPhone syncs with the following address book programs besides Outlook:

o   Windows Contact

o   Yahoo Address Book

o   Google Contacts

o   Mac Address Book

o   Microsoft Exchange

➢   Choose your contact program. Many of these programs require you to click the Configure button and then add your id and password. You can manage only one contact list at a time if you use a PC. Using a Mac system, you can manage multiple contact lists. If you use a business Microsoft Exchange account, you can only use this account. All of your personal contact and calendar data are deleted if you use Microsoft Exchange.

➢   Next you need to signify whether you want to sync all contacts or just certain selected groups. If you click Selected Groups, you have to choose who you want to sync and you need to stipulate where you want the other contacts to be saved.

Just the Facts About Syncing Calendars

Sync Calendars With determines how you will sync your calendars. As with contacts, the default is set to Outlook. You can choose All Calendars or Selected Calendars. You can also choose how many days to backup. Macs sync with iCal and can sync with multiple applications. PCs sync with Outlook and can sync only one application at a time.

Figure 3-5: iTunes Sync Info Page

Just the Facts About Syncing Mail Accounts

Sync Mail Accounts From lets you sync your e-mail account settings. You can synch all of your e-mail accounts or individual accounts. Syncing mail accounts syncs your account settings, but not your passwords or messages. You must save your passwords through settings on your iPhone, which were discussed earlier in Chapter 2.

Programs that sync with this are for Mac Mail and Microsoft Exchange and for PC Outlook or Outlook Express. Changes made to your e-mail accounts on your iPhone will not be synchronized back to the e-mail account on your computer.

Figure 3-6: iTunes Sync Info Page for Email

The Other option gives you two choices: Sync Bookmarks With and Sync Notes With. Bookmarks sync with Mac Safari, PC Internet Explorer, or Safari. Notes sync with Mac Apple Mail or PC Outlook (see Figure 3-6).

Advanced gives you the option to replace any of the following items on your iPhone from your computer:

➢ Contacts
➢ Calendars
➢ Mail Accounts
➢ Bookmarks
➢ Notes

If you get anything messed up on your iPhone, you can correct it by erasing the information on your iPhone by replacing it with what is on your computer. You can check one or all of these options to fix your iPhone. (see Figure 3-6). After the box(es) are checked, your information is exchanged on the next sync.

Just the Facts About Syncing Media

Ringtones, music, movies, TV shows, podcasts, videos, and books are synced only from your computer to your iPhone. If you delete any of these items from your iPhone, they are not deleted from your computer the next time you sync. However, if you purchase items and download them directly to your iPhone, they are synced to your computer the next time you sync.

Ringtones can be synced by checking Sync Ringtones in the Ringtones window. If you have customized ringtones in your iTunes, you can choose either all ringtones or individual ringtones by checking the box next to your choice.

Music can be transferred to your iPhone by clicking the Sync Music check box in the Music window.

Figure 3-7: iTunes Sync Music Page

You can choose Entire Library, Selected Playlist, Include Music Videos, and/or Include Voice Memos. If you choose Selected Playlist, a drop-down list shows Playlists, Artists, and Genres. You can then click next to the items that you want to download to your phone.

If you have more songs in your iTunes library than storage space on your iPhone, you receive an error message telling you that your phone cannot be synced because there is not enough space left on your phone to do so. You need to limit how many songs, videos, or voice memos you load onto your phone.

To know how much space you have used, you can look at the meter at the bottom of any of the information pages. As you can see at the bottom of Figure 3-7, the meter tells you how much of your iPhone storage space has been used.

Movies can be downloaded to your phone by selecting the Sync Movies check box, as shown in Figure 3-8. Next choose whether you want to automatically include all movies or selected movies. If you choose selected movies, check the boxes next to the ones that you want to download to your phone. They are added to your phone when you sync.

Figure 3-8: iTunes Sync Movies Page

TV Shows can be downloaded by checking the Sync TV Shows box on the TV Shows window, as shown in Figure 3-9. Then you can choose to include all watched or selected shows from the menus. You can choose shows and/or episodes by checking the boxes next to your choices.

Figure 3-9: iTunes Sync TV Shows Page

Podcasts, iTunes U, and Books can be downloaded by selecting the appropriate pages and checking your selections. You need the iBooks app from the App Store to download eBooks to your iPhone. I discuss apps further in Chapter 5.

Photos can be synced from your phone or your computer:

1. In the Photos window, choose **Sync Photos From**.
2. Select all folders or selected folders, as shown in Figure 3-10.
3. Check the box next to the items that you want to sync to your iPhone. You can sync your photos with any folder on your computer that contains images.

Figure 3-10: iTunes Sync Photos Page

If you have taken any pictures with your iPhone since your last sync, at the beginning of your next sync session, a window launches asking whether you want to sync your photos (see Figure 3-1). You can choose to download the new photos to your computer or not. If you choose to save them, you are asked to set up a folder where the pictures will be downloaded.

Remember to sync your iPhone often to make sure that you never lose any of your important data. If you lose your data, unless it is on your computer, it is gone forever. Save yourself a lot of headaches by taking the few minutes to sync your data.

Chapter 4

Using Key Features

Now that you have had a chance to set up your new Verizon iPhone, it's time to learn more about the special features that come with this phone. Hopefully, you have played with your phone and learned some of the features by trial and error. In this chapter, you are going to take a detailed walk through those bells and whistles that make your iPhone more than just a phone.

Organizing Your iPhone

You have all of these icons, and you really wish you could put them in a different order or organize them in a different way. You are going to learn how you can do this!

Moving the Icons

If you do not like the way the icons are arranged in the list, it is very easy to move them around. To move the icons around in the list order, complete the following steps:

1. Place your finger over the icon that you want to move.
2. When the icons start to wiggle, drag the icon that you want to move to the place in the list where you want to move it.
3. The other icons rearrange to fill in the gap.
4. When you are finished moving the icons around, push the Home button; the icons once again are locked in place.

Setting up Folders

Folders are a way to organize all of the icons on your iPhone. You can put like icons in the same grouping or folder so that you can find the various applications a lot easier. If you do not set up folders, you have to search all of the pages on your phone to find the one that you are looking for.

This process can be a real pain as your iPhone can grow to as many as 11 pages. Instead you can set up a folder that can hold up to 12 icons. To set up a folder, complete the following steps:

1. Place your finger on the icon that you want to move.
2. When the icons start to wiggle, move one icon on top of the other.
3. A window opens, and both icons automatically are placed in this window, as shown in Figure 4-1.

Figure 4-1: Icon Folder Window

4. If these are the icons that you want grouped together, push your Home button twice, and your new folder is created. Your icons are locked in place.
5. To add icons to your folder, drag icons over the folder. They are automatically added to that folder.
6. Again, push the Home button, and the icon is saved into the folder.

If you decide that you do not want an icon in a particular folder, do the following:

1. Tap on a folder, and it will open.
2. Put your finger on the icon until the icons start to wiggle.
3. Drag the icon out of the folder that it is sitting in.
4. Push the Home button, and the icons are locked in place. The icon once again is sitting on the Home page.

Apple automatically names the folders based on what icons are in them. To change the title of the folder, complete the following steps:

1. Place your finger over one of the icons until they start to wiggle.
2. The title is placed in a text box, as shown in Figure 4-2.
3. Tap the X on the right side of the text box.
4. The keyboard comes up so that you can type a new name in the text box.
5. When you are finished, double push on the Home button, and the folder locks in place with the new name.

Figure 4-2: Folder Name Change Window

You can change the order of the icons in the folder by moving them the same way that you would move your icons on the Home page: You tap on the folder, put your finger on the icon, and drag it to where you want it in the list.

When you have completed the changes, double push on the Home button, and all of the icons are locked in place. You can have as many as 180 folders on your phone with a maximum of 12 icons in each folder. You can make a folder disappear by taking all of the icons out of it. When you do this, the folder automatically disappears.

Special Screen Functions

Your screen has a couple of hidden functions that are important to know about. If you are on your Home page and flick it to the right, a search window opens along with a keyboard. Use this search to find something quickly on your iPhone.

A good example of when to use this is if you want to find your photos and you can't remember what folder you put them in. Follow these steps:

1. Type **Angry** in the search text box. Every instance in which angry shows up is seen in the space between the search text box and the keyboard, as shown in Figure 4-3.

Figure 4-3: iPhone Search Window

You then can choose the item that you are looking for by tapping on the item.

2. To clear the search, tap on the **X** on the right side of the text box and the search goes away.
3. Push the Home button, and the Search page goes away.
4. You can also flick the page to the left, and the Home page returns.

5. Another way to get to the search is to push the Home button one time. Pushing on the Home button one time makes the Search window appear and disappear.

Another handy tool is the Clean Memory window. Every time you open an icon, it is placed in the memory of your phone. The icons continue to utilize memory until you clean them out of memory.

Why is this important? The more memory being used on your phone, the slower your phone works and the faster you use up your battery. To clean out the memory, complete the following steps:

1. Double push the Home button.
2. A window opens that has all of the icons that you have used.
3. Place your finger on one of the icons until the red circle with a minus sign opens.
4. Tap the red circle, as shown in Figure 4-4, and the icon disappears.

Figure 4-4: Clear memory by deleting icons.

It is important to note that removing the icons from memory will not remove them from your phone.

Using Your iPhone to Take Pictures

One of the most attractive features of your iPhone are the excellent digital cameras. Actually two cameras are in your iPhone. The one on the back is a 5-megapixel autofocus camera, and the one on the front is a VGA camera. The iPhone has a built-in LED flash.

To take a basic picture, perform the following steps:

1. Tap the camera app.
2. Give the app a minute to open up and reveal your picture area.
3. Make sure the Picture/Video button on the bottom right side of the camera screen toolbar is pointed toward camera and not video.
4. Aim the camera where you want to take a picture.
5. When you are happy with the picture, tap the camera app at the bottom of the page, as shown in Figure 4-5, to snap the picture.
6. Be sure to tap the camera icon and not push the Home button to take the picture.
7. Be sure to remain still as a momentary shutter lag occurs to avoid taking blurry pictures.

Figure 4-5: iPhone Camera Screen

Turning your camera sideways allows you to take pictures in a landscape view instead of portrait view. Your camera shows a focus square before you snap the picture in the middle of the screen. This square shows you what the focus of the picture is. If you prefer to focus on something rather than where the initial square is, tap the area where you would prefer to focus, and it becomes the new area of focus.

Changing the Camera in Use

If you would prefer to use the front camera instead of the back, tap on the camera roll in the upper right side of the Camera screen. This is the camera with the pointing arrows. Now instead of taking a picture of what is in front of you, you can take a picture of yourself. If you tap this by mistake or you are ready to take pictures in the original view, tap the camera roll again, and the main camera again is in use.

Using the Zoom Feature

Your iPhone camera has a zoom that will zoom to up to five times. The zoom is a slider bar that you move to the right and left to move your image closer or farther away. To get to the slider, tap the screen, and it automatically appears, as shown in Figure 4-6. The slider only stays on-screen for a few seconds if you are not using it, so be sure to put your finger on it right away if you want to zoom in on something.

Figure 4-6: iPhone Zoom Slider Screen

How to Use the Flash

Your iPhone comes equipped with a flash. It is available only on the back facing camera. The flash symbol is located in the upper left side of the camera screen. The camera comes set in the auto position. The iPhone also has an on and off switch. To access the On and Off buttons, tap Auto in the upper left corner of the screen. Then you can tap the On or Off button to turn the flash on or off.

Managing Your Photos

You can see the pictures on your iPhone in two different ways. The first way is by tapping the box on the left side of the camera toolbar. The other way is by tapping the Photos icon on the main page of your iPhone. In either case, your pictures appear on your phone in small thumbnails. Tapping one of the thumbnails expands the picture to fill the screen. When a picture is the size of the screen, you can always flick the pictures side to side to flip through all of the pictures in the current album.

Downloading Photos to Your iPhone from Your Computer

Photos can be downloaded from your computer using PhotoShop Elements 3.0 or later, Aperture or iPhoto version 4.02 or later, or from any photo folders on your computer. You can copy only photos from one computer at a time. To copy photos from a second computer, you have to delete any photo that you copied from the first computer.

To download photos to your camera, copy the photos that you want to a folder and sync the folder to your iPhone. You need to first set up the destination folder on your iPhone. The folder syncs to your computer and then is copied back to your phone on the next sync.

Deleting Photos

Let's say that you're not happy with one or more of your photos, and you want to delete them. Pictures taken with your iPhone are very easy to delete, by following these steps:

1. Tap the thumbnail.
2. Tap the picture.
3. Tap the trash can in the lower right corner (see Figure 4-7).

Figure 4-7: iPhone Photo Screen

4. Another window opens, as shown in Figure 4-8. Tap **Delete Photo** to complete the delete or tap **Cancel** to not delete the photo.

Figure 4-8: iPhone Photo Delete Screen

If the photo is one that you downloaded to your iPhone from your computer, you need to delete it from the folder on your computer. The next time you sync your phone, the picture is deleted on your phone.

You can delete a group of photos by using the thumbnail view. Tap the arrow in the lower left side of the photo toolbar. Next tap the photos that you want to use, and you can either share, copy, print, or delete. On the upper right side of the photo screen is Cancel. If you change your mind about doing any of the listed actions, you can always cancel out of this screen.

Playing Slideshows

On the photo screen toolbar is a right-facing arrow. If you tap that arrow, your slideshow begins. Tap it again, and it end. Under Settings are four changes that you can make to a slideshow.

➢ Play each slide for either 2, 3, 5, 10, or 20 seconds.
➢ Transition is the image that you see between slides. Your choices are Cube, Dissolve, Ripple, Wipe Across, and Wipe Down. You have to try them to see which one you prefer.
➢ If turned on, Repeat will have your photos continue endlessly in Slideshow mode. If Repeat is turned off, it shows each slide one time.
➢ Use Shuffle to randomly show your photos in the slideshow when it is turned on.

You can also add music to your slideshow by starting a song on the iPod on your phone. (I talk more about using the iPod later in this chapter.) After starting your iPod, go back to photos and launch your slideshow.

Other Ways to Use Your Photos

When you tap the lower left corner of the photo toolbar, you get a selection of five other things that you can do with your pictures:

➢ E-mail: Tapping E-mail Photo takes you to an outgoing e-mail message where you can fill in to whom you want to e-mail the photo. And send your photo off like any other e-mail.
➢ MMS: Works just like e-mail. It is just another form of e-mail.
➢ Use As Wallpaper: Tap this to reset the wallpaper of your phone. Choose a picture, size the picture, and then tap Set Photo. You now have a new wallpaper picture.
➢ Assign To Contact: You can assign a picture to someone on your contact list. This will give you a quick reference of who is call you. To assign to a contact, follow these steps:

1. Find the contact to whom you want to assign the picture.
2. Hold and drag the photo to the photo square on the Contact page. You can size it to fit the square.
3. Tap **Set Photo,** and your photo is set.

You can also set a photo for a contact from the contacts list. To do this

1. Tap **Contact**.
2. Tap **Edit**.
3. Tap **Add Photos**.
4. A screen comes up prompting you to either Take Photo, Choose Photo, or Cancel, as shown in Figure 4-9.

   ➢ Taking a photo accesses the camera so that you can take the person's photo and add it to her contact page.

   ➢ You can choose a photo you already have taken by tapping Choose Photo. You are taken to your photos to choose a photo. Tap the photo that you want to use. Then tap it in place to set it.

Figure 4-9: iPhone Photo Select Screen

You can change the contact picture by tapping Contacts. Find the person whose photo you want to change. Tap Edit and then you can change the photo by following the preceding steps.

Taking a Screen Shot

You can take a picture of the various screens of your phone; for example, if you want to take a picture of a map or a website for later reference.

To take a picture of your screen, press the On/Off button while pressing the Home button. Release both buttons, and the picture is snapped. It only takes a couple of seconds of pressing both to take a picture of your screen.

Shooting, Downloading, and Watching Videos

Now that you are a pro at taking pictures with your iPhone, it is time to advance to taking videos. Not only can you shoot videos, you can download and watch TV shows, movies, and YouTube videos.

Shooting Videos

You can shoot videos in either portrait or landscape mode. To shoot a video, follow these steps:

1. Tap the **Camera** icon.
2. Tap the **Video** icon in the lower right side of the camera toolbar. The line moves from the camera to the video camera, as shown in Figure 4-10.

3. Tap the red Record button, which is just above the Home button. The red button blinks, and you see a counter timing the length of your video.
4. When your video is complete, tap the red Record button to stop recording. Your video saves to your camera roll.

You can tap the flash icon to have more lighting for your video. As in taking pictures, this can be found in the upper left corner of your video screen. You can also use either camera (the back or front cameras) to shoot video. Only one camera can be used at a time to record video.

Figure 4-10: iPhone Video Screen

Editing Videos

You can do some minor editing of videos using your iPhone. To edit your video:

1. Tap your video to display the on-screen controls, as shown in Figure 4-11.
2. Drag the start and end points along the timeline to select the video you want to keep.
3. Tap **Trim** located in the upper right corner of the video screen.
4. Last, you need to decide what to do with the video that you just trimmed. You can either save it as a new video, cancel to undo the changes, or delete to get rid of the trimmed piece of video.

Figure 4-11: iPhone Video Editing Screen

If you want a more sophisticated editing software, there is an app for that! (I discuss apps in Chapter 5.) You can download a video editing app. I understand several very nice programs exist for video editing for around $5. After you have finished editing your videos, you can play it. It can also be e-mailed or sent to YouTube.

Just like e-mailing photos, tap the arrow in the lower left corner of the video toolbar and the options to E-mail, E-mail MMS, or Send To YouTube appears. Make your choice by tapping the option that you would like to use. The E-mail or MMS option opens up into an E-mail screen where you can fill in to whom you want to send the e-mail.

YouTube is a little different as you need to have a YouTube account. You are prompted to add your YouTube log on and password. After you are logged on, you need to complete a form to be sent to YouTube along with your video. After you complete the form, you can send your video on to YouTube.

Downloading and Watching Movies, TV Shows, Music Videos, Podcasts, and YouTube

Movies and TV shows or music videos can be purchased or rented through iTunes. Open your iTunes icon and tap Videos. You can choose between movies, TV shows, and music videos. There are also reviews and previews. You have 30 days to watch a rented video. After you have downloaded your movie, TV show, or music video tap the iPod icon, tap Videos, and tap the item that you want to watch. (I talk more about using the iPod in the next section.)

A wide variety of podcasts is available on any topic on iTunes. Tap the iTunes icon, tap More. A menu of choices comes up. Podcasts are one of those choices. You can choose between

What's Hot, Top Ten, and Categories. Each choice has a rating. Most Podcasts are free to download. After you choose and download your Podcast, you need to open the iPod icon to watch the Podcast.

YouTube has so many choices, it has its own icon. Just tap on the icon and choose Away. You can watch all videos from this application. You do not need to go to your iPod to watch YouTube videos. Other choices are Featured, Most Watched, Search For A Topic, Favorites, or More.

Just the Facts About Using the iPod

To use the iPod, tap the iPod icon on the Home page. When the iPod opens up, you have several choices:

➢ Playlist, which is where your playlist resides after you download one. The Playlist option plays a single selection or shuffles to play random songs from throughout your playlist.

➢ Artist gives you an alphabetical list of all of the artists in your playlist. You can flick up and down the list to find the artist to whom you want to listen. A Search option is available at the top of the page that you can use to search for a particular artist. If you have more than one album for an artist, when his or her name comes up, so will a list of the albums that are on your phone by that artist. If you have single songs instead of the entire album, the single songs list instead of the album. Tap a song, and it begins to play.

➢ Songs gives you the list of songs on your iPod. You can flick through the list or use the search engine to search for a title. When you find the song that you are looking for, tap it to play it.

➤ Cover Flow is a special program that shows and lets you choose music by tapping the album cover. Simply turn your iPhone sideways, and Cover Flow automatically loads any album covers that are on your phone. Tap the album cover, and the songs begin to play.

➤ Videos gives you the list of videos on your iPod. You can flick through the list or use the search engine to search for a title. When you find the video that you want to play, tap it to play it.

➤ More allows you more search choices by tapping albums, audiobooks, compilations, composers, genres, iTunes U, or podcasts.

You can change the options in the toolbar. Say you want to change artists to albums. To change out the buttons, follow these steps:

1. Tap the **More** button.
2. Tap the **Edit** button located in the upper-left corner of the screen.
3. Drag any button on the screen to the button at the bottom of the screen that you want to replace. (In the earlier example, you could drag album to artist.)
4. You can also rearrange the buttons in the iPod toolbar while in this view.
5. When you have everything the way you want it, tap **Done**, and everything will be set.

When you are finished with the changes, any item that was changed will now show under the More button.

Just the Facts About iPod Settings

Under Settings are some preferences you can set to customize your iPod.

➢ Shake To Shuffle can be set to On or Off. In the on position, all you have to do is shake your phone to change the song that is playing. To change this setting:
1. Tap the **Settings** icon.
2. Tap **iPod**.
3. Tap the **Shake To Shuffle** button to turn the feature on or off.

➢ Sound Check when in the on position, makes it so each song plays at the same sound level. That way one song cannot blast you out because it is recorded at a louder volume than the song that played before it.

➢ EQ stands for equalizer. Equalizers increase or decrease the levels of specific sound enhancers. The iPhone has more than a dozen equalizer presets, which are tailored to a specific type of music. The best way to find out whether you want to use equalization is to listen to music while trying out the different settings. Start listening to a song. While listening, follow these steps:
1. Tap the **Settings** icon.
2. Tap **iPod**.
3. Tap **EQ**.
4. Tap the various presets to see how they change the song to which you are listening.
5. When you find a preset that you like, press the **Home** button, and the change is complete.
6. If you do not want to use any of the presets, tap **Off**, and no changes take place.

➤ Volume Limit sets the loudest listening level for audio or video. To set the limits:
1. Tap the **Settings** icon.
2. Tap **iPod**.
3. Tap **Volume Limit**.
4. Drag the slider to adjust the volume level to where you like it.
5. You can tap **Lock Volume Limit** to set a password so that others cannot change your volume settings.
➤ Turning Lyrics & Podcast Info on or off allows you to insert the lyrics or podcast info to your songs or podcasts when in the on position.

Making a Playlist

You can create a playlist on your iPhone in two ways. The first way is to create one in iTunes on your computer and sync that to your iPhone. The other is to make a playlist directly to your iPhone:

1. Tap the **iPod** icon.
2. Tap the **Playlist** button.
3. Tap **Add Play**list.
4. Type a name for your playlist .
5. Tap **Save**.
6. Tap the **+** next to a song name to add the song to your playlist.
7. Tap the **Done** button in the upper right corner.

If you create a playlist on your iPhone and sync it with your computer, the playlist saves to your computer as well. You have this playlist on your phone unless you delete it. To delete a playlist, select the playlist's name in the source list and then press Delete.

You can also edit your playlist on your iPhone:

1. Tap the **Playlist** button at the bottom of the iPod screen.
2. Tap the playlist that you want to edit.
3. At the top of the screen you see Edit, Clear, or Delete: Tap **Clear** to clear all of the songs out of the playlist; tap **Delete** to delete the playlist; tap **Edit** to do the following things:
   ➢ To move a song in the playlist, tap the icon with three gray bars on the right side of each song. Drag the song up or down depending on where you want to move the song.
   ➢ To add songs, tap the + button in the upper left corner.
   ➢ To delete a song from the playlist, tap the – to the left of the song name. This action only deletes the song from the playlist, not from your phone.

When you have completed your changes, tap Done, and the changes are locked in place.

Setting a Sleep Timer

Your iPhone has a sleep timer so that you can go to sleep listening to music, but do not have to use up your entire battery before it turns off. To set the Sleep Timer:

1. Tap the **Utilities folder.**
2. Tap the **Clock**.
3. In the lower right corner, tap the **Timer** button.

4. Set the number of hours and minutes you want the iPod to play and then tap the **When Timer Ends** button.
5. Flick to the bottom of the list and tap on **Sleep iPod**.
6. Tap the **Set** button in the upper right corner.
7. Tap the **Start** button.

If you have music already playing on the iPod, you are finished. If you are not playing any music, go into your iPod and choose what you want to play while you are going to sleep.

Using Voice Control to Run Your iPod

You can use your Voice Control option to do some things with your iPod. To use Voice Control, press the Home button, and the Voice Control option opens up. When Voice Control is open, you can do one of the following:

➢ To play an album, artist, or playlist, say "Play Album, Play Artist, or Play Playlist" and the name of the album, artist, or playlist you want to play. This command will work at any time other than when you are on a phone call or using FaceTime.
➢ To shuffle the current playlist, say "Shuffle."
➢ To find out more about the song that is playing, ask what you want to know. An example would be "What is the name of this song?" This works only while playing a song.

Just as with buying movies, TV shows, and music videos from iTunes, you also can buy and download music.

Tap the iTunes icon and tap the Music button. You can choose music by New Releases, Top Tens, or Genres. Reviews are available for the music. A price is given for the albums and also for the individual songs in the event that you are interested in only one or two songs on an album. Tap the Price box, and you are given the options to download the music.

Surfing the Web

Safari is the Web program used to navigate the web on your iPhone. You can gain web access through WI-FI or Bluetooth networks. To access the web, tap the Safari icon on your Home page toolbar. The first thing you need to do is choose your search engine from the list that is given.

The Web screen has many useful tools on it. The toolbar on the bottom of the screen has the left and right arrows in the bottom left corner, which take you to the previous and the next Web pages. The + in the middle of the toolbar helps you add bookmarks (sites that you use a lot or want to go back to without have to do a web search again). The Home screen takes a Web page symbol to your home page for easy access. For example, you can add Google to your Home page so that you can easily search the web.

Use Mail Link Page to link a web page to an e-mail message and send it to someone. The Open Book button is the home of your bookmarks. Last in this line is the Pages button.

When you have searched some pages, they stay open until you remove them. Pressing the Pages button takes you to all of the pages that are still open. If you no longer have a need for these pages, press on one of the pages. An x appears on the upper left corner. Tap this x, and the page closes. This action also saves some memory space by reducing the number of open web pages.

The toolbar on the top houses the address field where you can enter web addresses with which you are familiar. The partial circle with an arrow is the Reload Web Page button. To the right of the Reload button is the search engine that you chose, as shown in Figure 4-12.

Figure 4-12: iPhone Safari Screen

When you tap the Web Address box, the virtual keyboard appears. You will notice that on the bottom line is the .com key, as shown in Figure 4-13. Since most websites end in .com, Apple thought they would make all of our lives easier by adding this.

But what about those websites that end in .edu, .net, .us, and .org? Apple helped us out with those as well. If you tap .com and keep your finger there, a window opens with those suffixes as well. Do not lift your finger off of the .com, or the window closes. With your finger on the .com, slide it to the suffix that you prefer and press. The new suffix is used.

Figure 4-13: iPhone Web Search Screen

As you type letters, a list of websites matching the letters appears to help you quickly choose a website. You can choose one by flicking through the list or keep typing. You can also use the search engine to find a website.

To open a website, do the following:

1. Tap the **Safari** icon on the Home page toolbar.
2. Tap the **Address** field.
3. Begin typing the address that you want to use.

4.   Either choose a site from the list that comes up or continue typing until you have finished the address you are typing.

If you type an address by mistake, you can erase it by tapping the x on the right side of the address box. You can also use the x on the keyboard to erase a letter.

After you open a web page, the typing may seem a bit small. In order to read the page, you can double tap on the page to zoom in on a section of the site. Some other ways exist to make viewing your websites easier:

➤   Pinch the page. Use the pinch method to zoom on the open web page to see it better.
➤   Press down on a page and drag it in any direction to easily read what you want on the page.
➤   Turn your phone sideways to see the web page in the widescreen view. This should also help you read the page better.

Using Multiple Pages at the Same Time

You may want to look at more than one page at a time. It may be that you want to comparison shop before you leave one store's page. To do this:

1.   Tap the **Page** icon in the lower right side of the Safari toolbar. Your current page shrinks, and a New Page button appears.
2.   Tap the **New Page** button, and a blank page comes forward.
3.   Type the new address, and a second page opens. The number inside the Pages button lets you know how many pages are open.

4. When you are finished with all of your pages, if you tap on the **New Page** button, all open pages are available. Tap on them to access the Delete button to close all of the pages.

Using Bookmarks

To bookmark the site you are searching, tap the button just above the Home button on the Safari toolbar. Tap Add Bookmark, as shown in Figure 4-14.

Figure 4-14: iPhone Bookmark Screen

You begin to form a list of bookmarks. Editing your bookmarks may involve removing, changing, organizing or moving.:

To remove a bookmark, do the following:

1. Tap the **Bookmarks** icon.
2. Tap **Edit**.
3. Tap the red circle with the – next to the bookmark that you want to delete.
4. Tap **Delete**.
5. Tap **Done**, and your changes are made.

To change a bookmark name or location, do the following:

1. Tap **Edit**.
2. Tap the bookmark. The Edit Bookmark screen appears showing the name, address, and location of the bookmark.
3. Tap the fields you want to change.
4. In the Name field, tap the **X** in the gray circle and then use the keyboard to enter a new title.
5. In the location field, tap the **>** and scroll up or down the list until you find a new home for your bookmark.

To create a new folder for your bookmarks, do the following:

1. Tap **Edit**.
2. Tap the **New Folder** button.
3. Enter the name of the new folder.
4. Choose where to put the new folder.

To move a bookmark up or down on a list, do the following:

1. Tap **Edit**.
2. Drag the three bars to the right of the bookmark's name.

Saving Web Images

To save web images, go to the image. Press your finger on the image and tap the Save Image button. Saved images can be found in your camera roll. If you choose Copy instead of Save Image, you can copy the image to an e-mail, note, or other site.

Just the Facts About Safari Settings

➢ Search Engine allows you to change your search engine between Google, Yahoo, and Bing.
➢ When Autofill is turned on, Safari automatically fills out Web forms by using your personal contact information, usernames, and passwords.
➢ When Fraud Warning is turned on, it warns you when you inadvertently visit a fraudulent website.
➢ Java Script On or Off. Many websites use Java Script, but there have been security risks by allowing its use. Not using it may affect how some websites work.
➢ Block Pop-ups On or Off. Most Pop-ups are spam related; however, some websites have legitimate pop-up requirements. You probably should block pop-ups, but you can turn this option off when you reach sites where you need to allow pop-ups.
➢ The Accept Cookies options are Never, From Visited, or Always. The default is From Visited. You can change this if you prefer to always or never accept cookies.
➢ Clear History, Cookies, & Cache all clear out the little programs that websites use to either load faster or remember you when you visit them again. It is a good policy to clear these out every now and then. To clear them out, tap the item and then tap again to clear.
➢ Developer lets you turn on or off the Debug program. Debug automatically appears to help you resolve web page errors. It may or may not be helpful to the basic web surfer.

Sending a Text

The iPhone supports both SMS (Short Message Service) and MMS (Multimedia Messaging Service). SMS messages are short typed notes. Many services limit them to 160 characters. If you go over this limitation, your message could be broken into multiple messages when received. In order to send text messages, you need a text messaging plan through Verizon Wireless. An unlimited plan costs around $5 per month. If you have teenagers, this is a great plan to have.

Sending a Text Message

Tap the Message icon on the Home page toolbar. Tap the Pencil and Paper icon in the top right corner of the screen to start a new text message, as shown in Figure 4-15.

Figure 4-15: iPhone New Message Screen

The first field to complete is the To field. You can complete this in one of three ways. These ways are as follows:

➢ If the recipient isn't in your Contact list, type his or her cell phone number in this square.

➢ If the recipient is in your Contacts list, type the first few letters of the name. A list of matching contacts appears. Flick through the list and tap the name of the contact that you want to use. The more letters you type, the shorter the list will be. After you choose your contact, you can begin to type another name to add more than one person to your text.

➢ Tap the blue + icon on the right side of the To field to select a name from your Contacts list.

After your contacts are chosen, go to the text entry field to start typing your message. The text entry field is the oval box on top of the keyboard located between the Camera icon and the Send button. After you have finished your message, tap the Send button to send your message.

To add a picture or video to your outgoing message, tap the camera to the left of the text entry field box. You have the option to take a new picture or send a picture in your photo album. Tap the selection of your choice. Tap Choose. Your photo is inserted into your text message. Add whatever message you would like to add. When the message is ready, tap Send, and it will be sent.

To forward a message, tap the message that you want to forward. Tap Edit. Tap the message you want to forward. Tap Forward and fill in the contact information just as if you were sending an original message.

Receiving a Text Message

If you want to hear an alert sound when you receive a message, tap Settings. On the Home screen, tap Sounds, tap Text Tone, and then tap > to view the available tone choices.

When you have chosen the tone that you like, tap the tone, and a check mark appears next to that tone. Tap Sounds and then flick down the item list. Be sure to tap New Mail and Sent Mail to get the alerts. If these options are turned off, you do not receive the tone alerts.

If you receive a message when your phone is asleep, the name of the sender appears on the Unlock screen when you push the Home button to wake up your phone. If your phone is awake and unlocked when your message arrives, all or part of the message and the name of the sender appear on the screen in front of whatever is already there, along with Close and Reply buttons. To read or reply to the message, tap Reply.

To read or reply to a message after you have tapped the Close button, tap the Messages icon. The list of received messages appears. If the message that you are trying to find is not the first one, flick down the list to find the message that you are looking for. Tap the message. Read and reply away.

Deleting a Text Message

You can delete text messages in the following three ways:

➢ In the Message window, tap Edit and then tap the circle to the left of the message you want to delete. Tap Delete in the lower left corner of the screen. The message goes away. Tap

Messages, and you are taken back to the Message screen. The Message folder remains in the message list.

➢ In the Message window, tap Edit and then tap the circle to the left of the message you want to delete. Tap Clear All in the top left side of the screen. Again, all of the messages clear but the message folder remains in the message list.

➢ To delete the message folder and all of the contained messages, tap Edit. Tap the red circle to the left of the message. Tap Delete, and the message folder goes away.

Just the Facts About Texting

You can do a few other things of note with your text messages:

➢ Use the Search box on the top of the Message screen to search your texts for a word or phrase.

➢ You can use a Bluetooth keyboard to type your messages. You can find one of these at your local technology or office supply stores. The Apple Stores carry one made just for your iPhone.

➢ If you want to text someone in your Favorites or Recents list, tap the Phone icon on the Home screen and then tap Favorites or Recents. Tap the blue > icon to the right of a name or number. When the next screen opens with the person's information, flick down to Text Message. Tap this icon, and you are prompted to choose what number to text the person. After you have made this choice, you are taken to the Message screen where you can type your text message.

➢ To call or e-mail someone who has texted you or whom you texted, tap on the message. When the message opens up, tap Call. Tap Contacts and if you have an e-mail address listed, tap the e-mail address. You are taken to the e-mail screen to send an e-mail.

➢ You can add someone to your contacts list from a text by tapping the message to open it. Tap Contacts in the top right corner of the page. You need to tap Add To Current Contact, New Contact, or Cancel. You are taken to the Contacts screen where you can add the person's information into your phone.

➢ If a text message includes a web page, tap that web page address to be taken to the web page in Safari.

➢ If the message includes a phone number, tap the number to call the person.

➢ If the message includes an e-mail address, tap the address to send the person an e-mail.

➢ If the message includes an address, tap the address to open up the Maps program to find where they are located.

Making and Using Notes

Notes is a program that you can use to create notes for yourself. These notes can also be sent to someone through e-mail. To create a note, tap the Notes icon on your Home page and then tap the + in the upper right corner. The keyboard appears, and you can type your note. After you are finished writing your note, tap the Done button in the top right corner of the screen.

After you save your note, you can do the following:

➢ Tap the left or right arrow at the bottom of the page to look at the next or last note.

➢ Tap the Letter icon at the bottom of the page to e-mail the note using the Mail icon.
➢ Tap the Trash Can at the bottom of the page to delete the note.
➢ Tap the Notes button at the top left corner of the Notes screen to see a list of all of your notes. Tap a note to open it to read, edit, e-mail, or trash it.

E-mailing

In Chapter 2 I discussed how to automatically set up the e-mail account that is on your computer, how to get e-mails, read e-mails, thread e-mails, and send e-mails. In this chapter, I discuss special features of the e-mail icon.

The iPhone allows for multiple e-mail accounts. The nice thing about this is that you can have a personal and business account on your phone. If you want a business and personal account, you may need to check with the IT people at your place of business to make sure that your business account works with Microsoft Exchange. Your business account needs to work with this program to work correctly. The IT people at your work should be able to help you set up this account to properly interface with your work system.

Searching E-mails

To search through your e-mails to find that all important e-mail, tap the status bar to get to the Search box at the top of your e-mail inbox. Tap Search Inbox. The keyboard appears, and you can type a search item. All e-mails that match your search criteria appear. You can search anything that is in your To, From, or Subject fields.

If you are using Exchange, MobileMe, or certain IMAP-type e-mail accounts, you may be able to search messages that are stored on the e-mail provider's servers (the Cloud). When this is available, a Continue Search on Server message appears. Tap on this message to go to the Cloud to continue your search.

Reading Attachments

The iPhone reads most attachment files. If it does not read the file, you cannot tap the file to open it. To read an attachment, do the following:

1. Open the e-mail message with the attachment.
2. Tap the attachment.
3. Read the attachment.
4. Tap the **Message** button in the upper left corner of the screen to return to the message text.

Just the Facts About E-Mail

➢ To see all message recipients, tap Details to the right of the sender's name. All of the recipients come into view.
➢ To add an e-mail recipient or sender to your contacts, tap the name or e-mail address at the top of the message and then tap either Create New Contact or Add to Existing Contact.
➢ To mark a message as unread, tap Mark as Unread in the top of each message in blue with a blue dot to the left of the message. When the message is marked as unread, it goes back into the unread count on the Mail icon on the Home page. The mail also has a blue dot next to it in the inbox list to let you know that it is unread.
➢ To zoom in and out of a message, use pinch and spread.

➤ Links in a message can be accessed by tapping the link. Websites are opened into Safari. Telephone numbers open in the phone. Maps open in the Maps program. Day, date, or time can be tapped to create a Calendar Event. An e-mail address takes you to a blank e-mail message.

Saving a Message to Drafts

To save an e-mail message to drafts, do the following:

1. Start an e-mail message.
2. Tap the **Cancel** button in the upper right corner of the screen.
3. Tap the **Save Draft** button.

If you decide not to finish the draft, tap the draft, tap Edit, tap the circle to the left of the draft, and tap Delete. The draft is forever erased, so make sure that is really what you want to do.

Just the Facts About E-mail Settings

In order to review your e-mail settings, tap Settings, Mail, Contacts, Calendars.

➤ Under Mail you can set preferences for how many messages to show, how many lines to preview, font size, labels, Ask Before Deleting, Load Remote Images, Organize By Thread, Always Bcc Myself. Either tap the On/Off buttons to set your preference or use the > to look at and choose your various options.
➤ You can also enter a signature that will be attached to your outgoing e-mails. Open the Signature option. The keyboard

is there to type your signature. When you are finished, tap Done in the upper right corner of the screen.

➢ Another setting to check out is the sounds setting. If you want a sound to notify you of new mail or sent mail, tap Settings, Sounds and flick down to New Mail and Sent Mail. Tap the On/Off button depending on your preference.

➢ To set how often the iPhone checks for new messages, tap the Settings icon on the Home screen, tap Mail, Contacts, Calendars, and then tap Fetch New Data. If your e-mail program supports push and you have it turned on, new messages automatically are sent to your phone from the server. If your e-mail program does not support push, the Fetch settings apply. You need to check which Fetch option you prefer.

➢ You can stop or delete an e-mail account by tapping Settings, Mail, Contacts, Calendars, and the appropriate account. Tap the Account On/Off switch to turn the account off. To delete the account, flick to the bottom of the list and tap Delete Account.

➢ Also under account in the Settings section, Mail, Contacts, Calendars is an advanced setting. Under the advanced setting, you can set how often to remove deleted messages, whether to use SSL, Authentication, Delete from Server, and the Server Port to use. Set these per your server and personal preferences.

Using Mobile Me

MobileMe is a group of Internet services provided by Apple on a subscription basis for $100 per year. What does MobileMe do for you? It allows you to automatically update your iPhone with changes that are made to your computer, iPod, and iPad.

Another feature is that you can view and edit your vital information at **www.me.com**. Your phone is synced with a virtual server, so that you can access your information from any computer. MobileMe is one of the services often referred to as "Cloud" services.

At **www.me.com** you can set up and subscribe to your MobileMe account. A 60-day free trial is available for new users. You might try this to see whether you like it and would like to continue by purchasing the service. You can also get to www.me.com through the control panel on your computer. The Mobile Me icon was added to your control panel when you set up your iTunes account.

Making and Using Voice Memos

Apple includes a built-in digital voice recorder on your iPhone. This is a great program to record notes to yourself or a conversation, lecture, or special event. You can also download apps (to be discussed in the next chapter) ,which enhance this program.

Recording a Voice Memo

Tap the Utilities folder. Tap the Voice Memos icon. The screen opens up to a microphone, as shown in Figure 4-16.

Figure 4-16: iPhone Voice Memos Recording Screen

Tap the red Record button in the lower left side of the toolbar to start recording. The needle indicator located in the center of the toolbar moves to let you know that you are recording. You can pause the recording by again tapping the Record button. When paused, the needle still continues to move. When recording a clock appears on the top of the screen to let you know how long you have been recording. To adjust the recording level, move the microphone closer or farther from your mouth.

Listening to a Voice Memo

You can listen to your recording in one of two different ways:

➢ After recording the memo, tap on the button to the right of the audio level meter. A list of all your recordings pops up in chronological order. The most recent recording is on top. The recording on the top automatically starts playing.

➤ If you are going back later to listen to a recording, nothing starts automatically. Again tap the Play button. When the list pops up, tap the voice memo that you would like to hear.

The playhead can be dragged along the scrubber bar to move ahead to any point in the memo, as shown in Figure 4-17. This is the way to skip over information that you are no longer interested in hearing.

Figure 4-17: iPhone Voice Memos List Screen

If you do not hear anything after tapping Play, tap the Speaker button in the upper left corner of the screen. This should increase the volume of the speakers.

Trimming Recordings

To trim recordings to delete the information that you do not need, do the following:

1. Tap the right-pointing arrow next to the memo you want to trim.
2. Tap **Trim Memo**.
3. A blue tube that represents the recording appears. Drag the edges of this to adjust the start and end points of the memo.
4. Preview the edit before tapping the **Trim Voice Memo** button by tapping the **Play** button.

Your edits cannot be undone. Make sure that you are satisfied with your cuts before tapping Trim Voice Memo.

Labeling a Voice Memo

Your memos can be labeled to help you remember their contents. If you do not label the recordings, you simply have the date and time the recordings were made. This may be enough for you, but if not you can make a label by

1. Tapping the right-pointing arrow for the voice memo that you want to label.
2. Tap the right-pointing arrow in the box showing the date, time, and length of your recording.
3. Select a label for the list that appears.
4. Choose **Custom** to type your own label.

Sharing or Deleting a Voice Memo

To share a Voice Memo with someone else, do the following:

1. Tap **Share** from the Voice Memos list.
2. You are given the option to e-mail the voice memo.
3. You can delete any voice memo by tapping the memo and then tapping **Delete**.

Using Calendars

You can open the calendar by tapping the icon on the Home page. The icon changes daily to show the date and day of the week. The Calendar has three different views:

➢ List shows all current and future appointments in a list view. You can flick through the list to see all appointments.
➢ Day view gives you an hour-by-hour listing of the day's appointments.
➢ Month view gives you a monthly calendar with dots on the days that have appointments. In this view, the day's appointments are shown below the calendar.

Tap which view you prefer to open into the calendar.

Adding Calendar Entries

When you sync your phone, you should also sync any entries on the calendar on your computer. If, however, you set an appointment to your phone, which you later sync to your computer, you can add it by doing the following:

1. Tap the **Calendar** icon and then tap the **List**, **Day,** or **Month** button.
2. Tap the **+** button in the upper right corner of the screen.
3. Tap the **Title/Location** field and type as much information as you choose, as shown in Figure 4-18.

Figure 4-18: iPhone Add Event Screen

If your calendar entry has a start and end time

1. Tap the **Starts/End** field.
2. Use the number spinner at the bottom of the page to set the beginning and ending time, as shown in Figure 4-19.
3. Tap Done when you are finished.

Figure 4-19: iPhone Start & End Screen

102

Just the Facts About Calendar Entries

➢ To enter an all-day event, tap the All-day button so that On is showing. Then tap Done.

➢ If you are setting up a recurring entry, tap the Repeat window. Tap how often you want the event to repeat. Tap Done.

➢ Reminder alerts can also be set by tapping Alert. Next tap a time and then tap Done.

➢ Tap Calendar to send the entry to a particular calendar and then tap Done.

➢ If you want to enter notes about the appointment or event, tap Notes at the bottom of the Add Event screen. Type your note and then tap Done.

➢ To delete an entry, tap the entry, tap Edit, and then flick down to Delete Event.

Using Calculators

To access the calculator that comes with your iPhone, tap the Utilities folder and then tap Calculator. Your calculator has two different views and uses. In the portrait view, you have a basic calculator that does simple addition, subtraction, multiplication, and division problems, as shown in Figure 4-20. If you turn your phone to the landscape view, all of a sudden you have a scientific calculator that can perform difficult calculations, as shown in Figure 4-21.

Figure 4-20: iPhone Simple Calculator Screen

Figure 4-21: iPhone Scientific Calculator Screen

Setting Up Clocks

To access the clocks that come on your iPhone tap the Utilities folder and tap Clock. The clock has four different types of clocks:

➤ World Clock lets you see what the current time might be in numerous cities around the world, as shown in Figure 4-22. Tap the + button in the upper right corner and use the keyboard to start typing a city name. The minute you start typing letters, the iPhone lists the names of cities or countries that begin with those letters.

➤ You can create clocks for as many cities as you want, but only four clocks show at one time. You need to flick up and down the list to see the others. To remove a city, tap Edit, tap the red circle to the left of the city, and then tap Delete. You can rearrange the order of the list by tapping Edit. Then press your finger against the symbol with three horizontal lines to the right of the city you want to move. Drag the city to where you want it in the list.

Figure 4-22: iPhone World Clock Screen

➢ The alarm clock can be set by tapping the + in the upper right corner and then following these steps:

1. Choose the time of the alarm by rotating the wheel in the bottom half of the screen, as shown in Figure 4-23.
2. You can repeat the alarm on multiple days by tapping **Repeat** and then checking how often you want it to repeat. You can check every day or one day of the week.
3. Tap **Sound** to choose the sound that you want to wake up to.
4. Tap the **On/Off** button for the snooze to set it or not.
5. You can change the label for the alarm by tapping **Label** and typing a new title for it.
6. Tap **Save** when the alarm settings are where you want them to be.

A clock has been set and activated when the tiny status icon that looks like a clock appears in the upper right corner of the screen.

Figure 4-23: iPhone Alarm Clock Setting Screen

When your ring/silent switch is set to Silent, your iPhone does not ring, play alerts, or make iPod sounds. But it will play alarms from the Clock app. Make sure to turn your alarm off while sitting at a movie, play, or some other activity where it may sound.

Using the Stopwatch

To use the stopwatch function, simply tap Stopwatch and then tap Start. Tap Stop to end the timing and Reset to reset the stopwatch.

Using the Timer

To utilize the timer, tap Timer and then rotate the hour and minute wheels until the time you want is highlighted. Tap When Timer Ends to choose the ringtone that will let you know that the time is up. After you have the settings the way you want them, tap Start to begin. When the timer is finished, you hear the tone and then a Timer Done message appears on-screen. Tap OK to silence the ringtone.

Finding Your Way with Maps

To start the Map app, tap the Map icon on the Home page. To find a person, place, or thing, tap the Search field on the top of the Map screen and type an address or a place, as shown in Figure 4-24. A map comes up with pins for possible locations.

Tap the various pins to see the place that matches your search. If it is a place tap the > symbol, and it will open into a window that gives you the phone number, website, address, and the option to get directions, as shown in Figure 4-25. To get directions, type the beginning address; the ending is the place you searched. Tap Route to begin your route.

Figure 4-24: iPhone Map Direction Information Screen

You can also add it to contacts, share the location, or add it to bookmarks by tapping the appropriate box. You can also go back to the map. You can tap the pin to zoom in on the location. You can also use pinch and spread to zoom in on the map, eventually getting to the street level.

If you tap on the button that looks like a curling piece of paper in the lower right corner of the screen, the map rolls up to reveal some other options, as shown in Figure 4-26. You can replace the pin or hide traffic. You can also alter the view to show either the map, the Satellite view (aerial view), Hybrid (aerial view with the street names superimposed), or list (list of directions).

Figure 4-25: iPhone Map Location Information Screen

Figure 4-26: Other Map Options

109

In the Map view, you have the choice of transportation mode:

➢ Driving view
➢ Public transportation view
➢ Walking view

Choose your mode of transportation. Your map alters depending on the mode of transportation you choose, as shown in Figure 4-27.

Figure 4-27: iPhone Map View Screen

Using the Compass

To open the compass, tap the Utilities folder and then tap Compass. The compass opens and automatically starts working.

Tracking Stocks

To get stock information, tap the Stocks icon on your Home page. The default screen is the various stock indexes and the latest market trends over the past six months, as shown in Figure 4-28. You need to flick the list of the various indexes to see the entire list. If you flick the market trends side to side, you see the latest financial news headlines and the Dow Jones Industrial Averages.

Figure 4-28: iPhone Stock Screen

To customize the screen to show your own stocks, do the following:

1. Tap the **i** button in the bottom right corner of the initial Stocks screen.
2. Tap the **+** button in the top left corner of the Stocks screen.
3. Type the stock symbol or the name of the company, index, or fund.
4. Tap the **Search** button.
5. Tap the one you want to add.
6. Repeat steps 4 and 5 until you are through adding stocks, funds, and indexes.
7. Tap the **Done** button in the top right corner.

To delete a stock or index, do the following:

1. Tap the **i** button in the bottom right corner of the Stocks screen.
2. Tap the **–** button to the left of the stock's name.
3. Tap the **Delete** button that appears to the right of the stock's name.
4. Repeat Steps 2 and 3 until all unwanted stocks have been deleted.
5. Tap the **Done** button.

To change the order of the list, tap the i button and then drag the three horizontal lines to the right of the stock, fund, or index up or down to its new place in the list.

Finding Stock Information

To see the information about the stock that is in your list, tap its name to select it, and the lower portion of the screen shows additional information. You can look up more information about the stock by tapping the Y! in the lower left corner of the screen.

Tapping the Y! takes you to the Yahoo.com finance page for the particular stock that you are interested in.

Tracking Stock Charts

The chart shown in Figure 4-29 tells you the trends of the stock over 1d (one day), 1w (one week), 1m (one month), 3m (3 months), 6m (six months), 1y (one year), and 2y (two years). If you turn your phone sideways, the chart takes up the entire page.

Figure 4-29: iPhone Stock Charts Screen

In this view, you can learn three things about your stock:

➢ Touch any point to see the value of the stock for that day.
➢ Place two fingers on any two points in time and see the difference in values between those two days. See Figure 4-30.
➢ Flick the chart left or right to see the chart for another stock, fund, or index.

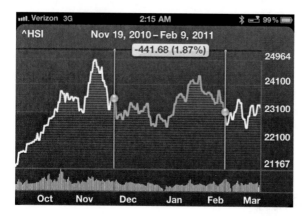

Figure 4-30: Tracking Stock Value

The Stock screen shows the change in stock prices in dollars. The dollars can be changed to show the stock value change by percentage or as the stock's capitalization. Tap the number next to any stock to change the display for all stocks. Green numbers are positive; red numbers are negative.

In the original portrait screen, tap the i button in the lower right corner of the screen. At the bottom of the next screen, you can tap percentage, price, or market cap to make your stock show in one of those views all of the time. After you are finished making your choice, tap Done, and the change is set.

Checking the Weather

I must admit that the weather app is one of my favorites (see Figure 4-31). Tap the Weather icon to get the current weather conditions for where you are plus the forecast for the week. During the day, the screen is blue. At night, it is dark purple.

Figure 4-31: iPhone Weather Screen

Tap the i button in the bottom right corner to add a city. On the City screen, tap the + in the upper left corner to type the name of the city and state or ZIP code that you would like to add. You can add as many cities as you would like.

To delete a city, do the following:

1. Tap the **i** button in the bottom right corner.

2. Tap the red circle to the left of the city.
3. Tap the **Delete** button that appears to the right of its name.

Tapping the i button allows you to choose between Fahrenheit and Celsius by tapping one or the other. It also allows you to change the order of the cities. The city at the top of the list is the default city on the weather screen. To move the cities in the list, tap the three lines to the right of the cities. The city can then be moved anywhere on the list.

When you have the list the way you want it, tap Done, and the list again is locked in place. If you have more than one city in the list, flick the screen to the left and right to see the various cities weather.

When you tap the Y! in the lower left corner of the Weather screen, you access the Yahoo weather page on the Internet. You can get more detailed weather information on this site.

Setting Restriction Controls

Restriction controls can be placed on any application on the iPhone. Many parents and businesses may want to limit how the phones are used. To set the controls, do the following:

1. Tap **Settings**.
2. Tap **General.**
3. Tap **Restrictions.**
4. Tap **Enable Restrictions**, as shown in Figure 4-32.
5. You are prompted to set a passcode of your choosing.
6. After setting the passcode, tap **Done**, and the passcode is set and the restrictions are now able to be set.

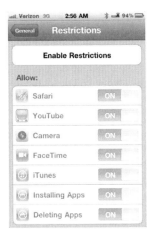

Figure 4-32: iPhone Restrictions Screen

You need to go down the list and set what restrictions you care to set. The first section is just yes/no prompts. Flick down to see the entire restrictions list. Under Allowed Content, tap on the > to see the various restriction options.

When you determine that the restrictions are no longer needed, tap Disable Restrictions. You are prompted to add the passcode. After the passcode is entered, you are taken back to the Restrictions screen. The Restrictions screen now has a gray tone, and you cannot tap any of the options.

Working in Airplane Mode

While on an airplane, you cannot use your iPhone, Internet, or e-mail. You can, however, use the iPod if you place your phone in Airplane mode. Airplane mode turns off those things that you cannot use on an airplane while allowing you to use those applications that you can use.

To get to Airplane mode, tap Settings. The first option in the settings list is Airplane Mode. After you land, you can go back to Settings and turn it off.

You have now learned the current functions of your new iPhone. As you can see, you can do so much with your new phone. About the only thing it cannot do is housework. There might, however, be an app (program) to give you helpful hints to make your housework lighter! In the next chapter you will learn all about apps.

Just the Facts About Key iPhone Features

➤ To move an icon, put your finger over it. When it begins to wiggle, move it where you want it and push the Home button to lock the icons in place.
➤ Folders can be made to hold up to 12 icons in order to better organize your Home pages.
➤ To set up a folder, place your finger on the icon. When it begins to wiggle, place it on top of the icon with which you want to group it. When finished push the Home button to lock all of the icons in place. You can now drag other icons to this new folder.
➤ A hidden search function exists on your Home page. Flick the Home page to the right or press the Home button once to make the search function come forward. You can now search for anything on your iPhone.
➤ Clean the memory to keep your phone working at maximum speed and using less battery power.
➤ The iPhone has a front and a back camera.

- ➢ To take a basic picture, tap the Phone icon. Aim the camera, point, and shoot your picture.
- ➢ You can take pictures in landscape or portrait.
- ➢ Tap the camera roll in the upper right corner to change between the various cameras.
- ➢ Tap the Camera icon to open the slider bar on the camera face to open the zoom.
- ➢ Tap the Photo icon on the left corner of the camera to see your pictures.
- ➢ Syncing your phone saves your pictures to your computer.
- ➢ To delete a bad picture, tap the Trash Can in the lower right corner. Next tap Delete, and the pictures are deleted.
- ➢ Tap the right-facing arrow on the photo screen toolbar to start the photo slideshow.
- ➢ Start the iPod and then start the slideshow to have music playing while showing your pictures.
- ➢ Photos can be added to your contacts so that you can see who is calling you.
- ➢ Open the camera, move the button in the lower right corner from photos to videos, point, and shoot to take videos.
- ➢ You can download movies, TV shows, music videos, podcasts, and YouTube videos from iTunes to your phone. Some cost; some are free.
- ➢ Tap the iPod icon to open the iPod. You need to download music or videos from your computer through iTunes or from iTunes on your phone. After the music or videos are downloaded, you can tap them to play them.
- ➢ Your iPod has a sleep timer so that you can go to sleep listening to music and not run your battery completely down.
- ➢ You can use Voice Control to run your iPod.
- ➢ Use Safari to surf the web.

➤ To send a text message, tap the Message icon, choose to whom you want to send a text, then type the message in the text box, and tap Send.

➤ Notes is a program used to make notes to yourself. Tap the Notes icon and then the + in the upper right corner to open up the Notes program and write a note to yourself.

➤ You can have more than one e-mail account.

➤ You can subscribe to MobileMe, which is a mobile server in the "Cloud" network. The service is $100 per year.

➤ Verbal notes, conversations, lectures, or special events can be recoded using Voice Memos.

➤ Appointments can be managed using Calendar. The Calendar has three different views: list, day or month. Choose the view that you like the best.

➤ Two calculators are on your iPhone. There is a basic calculator (in portrait mode) and a scientific calculator (in landscape mode).

➤ Your iPhone comes with four different clocks: a World Clock, a Stop Watch, a Timer, and an Alarm Clock.

➤ Tap the Maps icon to find a person, place, or thing.

➤ Daniel Boone had nothing on the iPhone, which has its own compass to help you find your way.

➤ Follow and chart your favorite stocks through the Stock icon.

➤ Use the Weather icon to find out what the weather is doing where you are or where you're going to be.

➤ Restrictions can be placed on just about anything on your iPhone.

➤ While flying, put your phone in Airplane mode to use those functions (such as the iPod) that are allowed on an airplane.

Chapter 5

There Certainly Is an App for That!

One of the most intriguing and fun things about the iPhone is all of the apps that you can download to your phone to do just about anything. I have truly been envious of my friends who had iPhones and were constantly showing me the latest things they could do on their iPhones. Now that I have a Verizon iPhone, I have explored the available apps and found some true favorites. I'm sure that your favorites will differ from mine, but it's always fun to explore.

What Are Apps?

Comedians talk about apps; they are mentioned in commercials; apps pop up just about everywhere, and they are not necessarily limited to the iPhone. So what are they really? Apps are computer programs that someone much smarter than I took the time to design, engineer, and put in computereze. You can go to the App Store icon on your phone or through iTunes on your computer and download these programs for anything from free to $200 plus for some of the more technical business applications.

As of this writing there are more than 200,000 different apps and more are coming online daily. Many apps are free or $.99. They cover the gamut of things from a flashlight to photo editors, a tip calculator to news shows, how to books to ones on travel information. One of the more common is as a game controller. It is interesting just to go to the App Store and look around and explore the various options. There are so many different choices it can become a bit overwhelming. This chapter should help you find your way through the maze of apps that are available.

Where to Find Apps

In order to use the App Store to download apps, you have to have an iTunes account. An icon on your iPhone Home page takes you to the App Store, the most common place to find apps. You can download apps from the App Store through your iPhone or through iTunes from your computer. I discussed how to set up your iTunes account in Chapter 1.

Figure 5-1: iTunes App Store

iTunes for the Computer

Open your iTunes account on your computer:

1. Click the iTunes icon on your desktop.
2. Click **Store** on the menu bar at the top left side of the iTunes page.
3. Click **Sign In**.
4. You need to put in your user name (e-mail account most likely) and the password that you set up when you opened the account.
5. Click **App Store**.

Have fun exploring! The iTunes App Store is broken into sections. At top are new apps, in the middle are is the apps considered to be hot, next are current events kind of apps, and last are staff suggestions. Information about the various apps is provided. On the right side is a listing of paid apps, free apps, top grossing apps, and an area to search for apps. A search box also is provided in the top right corner of the iTunes screen.

Start exploring by clicking an app, which takes you to more information about the app. You learn whether there is a price, what the app does, and how much space it uses on your phone. You can also go to the manufacturer's website for additional information.

The Internet is another source of detailed information about various apps. Do a Google search for iPhone apps, and you will find a number of websites that rate apps. These websites have a lot of information regarding the use and functionality of a myriad of apps.

The rating system for apps is very helpful. Somewhere on each app screen is the assigned rating. Tap the rating to get information about what the ratings mean. Ratings are as follows:

➢ 4+ contains no objectionable material.
➢ 9+ may contain mild or infrequent occurrences of cartoon, fantasy, or realistic violence; or infrequent or mild mature, suggestive, or horror-themed content that may not be suitable for children under the age of 9.
➢ 12+ may contain infrequent mild language; frequent or intense cartoon, fantasy, or realistic violence; mild or infrequent mature or suggestive themes; or simulated gambling that may not be suitable for children under the age of 12.
➢ 17+ may contain frequent and intense offensive language; frequent and intense cartoon, fantasy, or realistic violence; mature, frequent, and intense mature, suggestive, or horror-themed content; sexual content; nudity; depictions of alcohol, tobacco, or drugs that may not be suitable for children under the age of 17. You must be at least 17 years old to purchase games with this rating.

Downloading Apps to Your Computer

Downloading apps is easy. After you have done your research, and you find an app that you want, click the Buy App or Get App button on the app screen. You are prompted to log into your iTunes Store account. After the app has downloaded, it shows in the Apps section of your iTunes library.

The apps can be shown on your computer through your iTunes Home page in any of four different views. The Grid view shows the app icons in a series of lines. The Song List view gives you a list of apps one on top of the other. It shows the rating and some other technical information about the app. The Album List view gives you all of the information the Song List gives you, but it also shows the App icon.

Cover Flow works just like music Cover Flow. You see the app icons. You click one of them to bring it forward to get information about that particular app. The buttons you use to choose your favorite view are to the left of the search box on the iTunes Home page.

After you have downloaded your app(s), sync your phone, and the app(s) will appear on your phone.

Updating Apps to Your Computer

Software developers often update apps to fix a glitch or add new components. You can check for updates by clicking Check for Updates in the lower right corner of the iTunes screen.

Finding Apps on Your iPhone

Finding and downloading apps to your iPhone is very similar to downloading them on your computer. One requirement is that you have an Internet connection.

Tap the App Store icon on the Home page of your iPhone. When opening the App Store, you see five things at the bottom of the screen:

➢ Featured
➢ Categories
➢ Top 25
➢ Search
➢ Updates

Select Featured, and you will find New, What's Hot, and Genius. Genius is a program that suggests apps to you based on the apps that are already on your iPhone.

Select Categories for a list of app categories. Tap > to find more information about the category. Tap > again, and you receive a list of apps in that category along with the Top Paid, Top Free, and Release Date options. Tap one of these options to further narrow your choices. Twenty to twenty-five apps appear on each page. You need to flick the list to see all of the choices.

Select the Top 25 option also provides Top Paid, Top Free, and Top Grossing options. Again, you need to flick through the list to see the entire list.

Select Search to access the Search window. Type the name of the app you're looking for, and the list populates with your possible choices.

Tap an app to learn more about it. An information page downloads. Flick down the page to read about the app and what it does. As you continue down the page, you get to the ratings area. You can read ratings from users, get the rating I mentioned earlier, share the app with a friend (Tell a Friend), gift the app to a friend (Gift This App), or Report a Problem. You can also download from this screen by pressing the price of the app.

Figure 5-2: iPhone App Store Screen

Downloading Apps to Your iPhone

Tap the Cost box to start the download process. You are prompted to log into your iTunes account. After logging in, the App icon appears in the next free space on your Home page. A status bar lets you know the progress of the download.

When the download is complete, the name of the app appears under the App icon. If the App has a 17+ rating, you need to tap the OK button to indicate that you are 17 or older.

Next, sync your phone to your computer. Syncing saves the new app to your iTunes account on your computer. If your phone should happen to have a problem, you won't lose your app, as it will be stored on your computer and ready to sync again once the phone is working.

Updating Apps on Your iPhone

You can update your apps in two different ways. When you log into your iTunes account, you are prompted to look for updates. If you click OK to do this, any updates to your existing apps download. You can also tap Updates from the App Store screen. (Again you need to log into your iTunes account. And you will have to have Internet access.)

Deleting Apps on your iPhone

Apps can be deleted through your iPhone and on your computer. To delete an app, do the following:

➢ On your computer, right-click the app's icon and click Delete. You are prompted to be sure this is the action you want to take. After you agree to the deletion, the app is deleted.

➢ On your iPhone, place your finger over the app. When the icons start to wiggle, a black circle with an X appears in the upper left corner of the icon. Tap this circle, and you are prompted to be sure that you want to get rid of this app. Agree to the deletion, and the app is deleted.

You need to delete the app on both your computer and your iPhone. Otherwise, the undeleted copy from the other device loads again when you sync.

Organizing Your Apps

Just as with any other data stored in folders, you can organize your apps. If you have a bunch of games, you can put one game icon on top of the other to start a Games folder. Your iPhone names the folder, or you can create a custom label. You are prompted to change the name when the folder is made:

1. Tap the **X** next to the given folder name, and a typing block opens so that you can type your custom label.
2. Type the label you want.
3. Tap Done, and your new folder opens up with the original two icons. You can add up to 10 more icons into a folder.

Suggested Apps

As I explored the available apps, I discovered a few favorites and many others that were interesting or fun. Here are some of the apps I suggest you explore and a brief description of what they offer.

➢ **Angry Birds** is one of the most popular game apps ever created. It has several free sample programs and several that cost $.99. This game pits various birds with various powers against a series of obstacles. I must admit, I have become an addict of this game app.
➢ **Flashlight**, just as the name suggests, turns your iPhone into a flashlight. It is great for many uses. Your LED camera flash becomes a flashlight when the app is in the on position. This app is wonderful to have on a stormy night when your electricity goes out.
➢ **Zosh** is an valuable program for business use. If someone needs your signature, they can send the document to you. You can open it, sign it using Zosh, and resend the document to whomever needs it. The best news—this is a free program.

- ➢ **Amazon** enables you to buy and download books, shop their website, and do the same things you can do at Amazon.com on your computer.

- ➢ **Starbucks** offers the coffee shop's reward cards right on your phone. You can also put money on account so that you can stop in any Starbucks, scan your card, pick up your coffee, and go. Store locations can also be found using this app.

- ➢ The red box, **Netflix**, also has an app. You can order your movies right from your phone.

- ➢ **Bar Code Scanner** lets your fingers do the shopping. Take a picture of the bar code of an item you want to buy. This app shows you where you can find the cheapest version of the item you scanned. It costs $0.99 but is well worth the savings.

- ➢ **Cardstar** enables you to take pictures of all of your reward cards—grocery stores, convenience stores, and so on. Now, you can scan the cards right from your iPhone. You no longer have to dig through your purse, pockets, or wallet to find all of the cards or key chain fobs.

- ➢ **Evernote** is a note program. It goes well beyond the note program that comes with your iPhone. This app is listed as one of the top apps to have on Apple's Top 10 List. Images and recordings can be added to notes to make them much more helpful.

- ➢ **Skype** is the same phone program you might be familiar with from your computer that enables you to talk or text with other Skype users overseas without incurring international calling charges. This is another super free app.

- ➢ **Mobile Noter** is another free note app. It interfaces with Microsoft One Note, making it an optimum mobile office tool.

➤ **Grocery IQ** lets you put together a grocery list including current pricing. Set it up like the aisles in your grocery store and mark the items that you want to buy. Another super free program.

➤ The major drug store chain apps are also a great resource, including the **Walgreens** app. Through this app you can order pictures, prescriptions, find stores, shop, or set up your shopping list. Ordering prescriptions is a snap, with no telephone menus to navigate.

➤ **Movie Finder** is the movie planner extraordinaire. With this app, you can find a movie, read reviews, find a local theater where the film is showing, invite friends to come, send your friends info about the movie, and then write your own review after you have seen the film.

➤ **Instagram** is a photo program that enables you to take pictures, add filters, add labels and share with friends, Facebook, Twitter, and other social media. You can also add cool effects to your photos to turn them into works of art!

➤ You should also take a look at one of the app store starter kits. These starter kits provide a selection of free and paid apps including social media, travel, news, sports, hobbies, and many more. An easy way to add a wide variety of apps to your phone without spending a lot of time searching.

Part of the fun of the iPhone is exploring the many different types of apps. I'm sure you will find some great apps by just poking around at the app store. Millions of apps are out there — find a few that are crazy and kooky and interesting just to show your friends!

Just the Facts About Apps

➢ Apps are programs used to perform some sort of activity or to entertain.

➢ You can find apps through the iTunes program on your computer or through the App Store on your iPhone.

➢ A rating system lets you know what apps are age appropriate and gives the parameters of what you might find in an app.

➢ To download an app, log into your iTunes account, tap or click the Buy box, and the app begins to download.

➢ Apps can be updated through the iTunes program on your computer or through the App Store icon on your iPhone.

➢ When deleting an app, delete it from both your computer and your iPhone.

Chapter 6

Troubleshooting

Now that you've become an expert (or at least modestly proficient!) with your iPhone, what happens if you start having problems with the phone? Where do you turn? This chapter helps you answer those questions.

Just the Facts About Troubleshooting Your iPhone

Very few iPhone users report problems with their iPhones. Various consumer services such a Consumer Reports , JD Power and Computer World rate the iPhone very high on reliability. Change Wave Research recently did a study of 4,200 Smart Phone users. Their poll concluded that 74% of iPhone users were happy with their phones versus 43% of Blackberry users. But like all technology, problems may occur. Apple suggests the following first steps if your iPhone starts acting up:

1. Recharge your phone to its full charge.
2. If recharging does not help, try restarting it.
3. The next level of troubleshooting is to reset your phone. To reset your phone, press and hold the On/Off button at the top of your phone, while pressing and holding the Home button. When you see the Apple logo, you can release both buttons. Your data should not be affected by a restart.
4. The next thing to try is to remove some or all of the content on your phone. Try to remember what you were doing when

your phone froze. Whatever it was, try removing that data or app first. Remove data one section at a time. When you figure out what is causing your trouble, you can permanently remove it. If you cannot figure out what was causing the problem and you have removed all data and apps from your phone, try restoring the data and apps one at a time and see whether you can figure out the problem that way.

5. If none of the preceding steps work, you can next try resetting your settings and content. To do this, first reset all off your settings to the default setting that came on your phone. Resetting will not affect any of your data. If the resetting does not help, you can try **Erase all Content and Settings**. To find Erase all Content and Settings

   ➢ Tap Settings
   ➢ Tap General
   ➢ Tap Reset.
   ➢ Tap Erase all Content and Settings

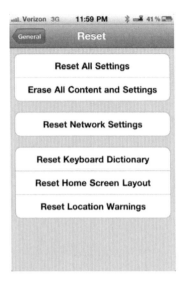

Figure 6-1: iPhone Reset Page

6. Next try restoring. Connect your phone to your computer as if you are going to sync it. Instead of syncing it, when your phone shows in the iTunes window, click **Restore** and follow the instructions on your computer.

See "Help! Nothing Is Working! Where to Next?" at the end of this chapter if you're still struggling.

Just the Facts About Troubleshooting Calls and Networks

If you are having trouble making or receiving phone calls or text messages, having Wi-Fi problems, or experiencing problems with your wireless carrier, try the following troubleshooting techniques:

➤ Check the cell signal icon in the upper left corner of your iPhone. If you do not have at least one or two bars, you cannot use the phone or messaging functions.
➤ Make sure that your phone is not in Airplane mode.
➤ Try moving around to see whether you get better reception in a different area.
➤ Much has been made about the antenna problems with the iPhone 4 when you grip the outside frame of the phone — sometimes called "The Grip of Death." RIM, the maker of the Blackberry claims that this is a design flaw in the iPhone due to the antenna wrapping around the outer bezel of the phone. Whether or not this is a design flaw with the iPhone, changing your grip or using a protective case seems to resolve most if not all problems with dropped calls.
➤ Turn on Airplane mode, wait 15 minutes, and then turn it off again. This resets your wireless and Wi-Fi connections.
➤ Restart your phone.
➤ Make sure that you SIM card is firmly seated. As many different types of iPhones are available, you might want to go to www.apple.com/support/iphone or your local Verizon Wireless store to get help with this.
➤ If nothing else helps, restore your phone as described in the previous section.

See "Help! Nothing Is Working! Where to Next?" at the end of this chapter if you're still struggling.

Just the Facts About Syncing, Computer, and iTunes Issues

If you are having trouble syncing your phone or your computer does not recognize your iPhone, try the following:

➢ Recharge your phone.
➢ Try a different USB port or a different cable.
➢ Restart your iPhone and try to sync again.
➢ Restart your computer.
➢ Reinstall iTunes. You can get the latest and greatest version at www.apple.com/itunes/download.

Help! Nothing Is Working! Where to Next?

Please do not give up and throw your iPhone against the wall — you really will have troubles then!

The first place to go is the Apple website. The URL is www.apple.com/support/iphone. You can choose your topic and see what Apple has to say about it. Apple also directs you to a users' forum. If you do not find your question in the users' forum, you can post a question and wait for others to answer. I found this website very helpful when I first got my iPhone.

You can also try Googling your question. After adding your question to the Google query box, many sites will pop up that you might find useful. I was having a problem with outgoing email due to using a cable Internet carrier. I googled this problem and many sites with potential help were listed.

If you still do not have any answers, try the technology people at your Verizon Wireless store. If they cannot help you, try Apple Cares at 1-800-694-7466. Your final option is to send your phone to Apple to have them look at it.

You have a year from the original purchase date of your iPhone to purchase a two-year extended warranty program through Apple for $69.

If you send your phone to Apple for repair, note the following:

➢ Your iPhone is erased during its repair. You lose any information that is not synced prior to repair.
➢ Remove any accessories such as a case or screen protector.
➢ Remove the SIM card and put it away for safe keeping. Be sure not to lose it. Apple does not guarantee that your SIM card will be returned after repair.
➢ If you have a warranty or extended warranty program, you can take your iPhone to an Apple store. For $29, they will replace it with a new iPhone. You can still get this service without the warranty program, but it is a lot more expensive. Double-check the cost from the Apple website at www.apple.com/support/iphone/service/exchange   as pricing frequently changes. The great advantage of this program is that you move the SIM card to the new phone and all you have to do is sync the new iPhone with your iTunes account. You then are good to go.

It is my hope that you do not have to use any of the suggestions in this chapter. Enjoy your new iPhone—it is such a great source of information, fun, and entertainment. I hope that you are finding yours as wonderful and useful as I am finding mine.